THE UNIVERSITY OF HULL

CENTRE FOR SOUTH-EAST ASIAN STUDIES

Occasional Paper No. 21

IRRIGATION AND RICE CULTIVATION IN WEST MALAYSIA

by

Amarjit Kaur

© Centre for South-East Asian Studies, 1992

ISBN: 0-85958-594-8

ISSN: 0269-1779

Excavation of Irrigation Canal. Removing a tree stump. Pulau Besar Irrigation Scheme, 1936.

Photograph by Alfred Wynne of the Drainage and Irrigation Dept. of the Government of Malaya. Mr. Wynne became a prisoner of war and died in captivity in Kuala Lumpur (Bateson Archive 14, Centre for South-East Asian Studies, University of Hull).

One of the most significant changes affecting the padi sector in Malaysia since Independence has been the introduction and spread of high-yielding varieties of rice with the accompanying technologies in the form of irrigation facilities and chemical inputs and pesticides. Additionally, through price support programmes, input subsidies and delivery systems, and the provision of extension services, the Malaysian government has intervened in the agricultural market to subsidise padi production. These measures have led to increased food production, lessened dependence on imported rice and have drawn the peasant economy into the national economic system. While several studies have been conducted of specific matters such as national rice policy, price support programmes and tenancy,[1] no in-depth study has been made of the role of irrigation in rice cultivation. This paper contributes to that effort by reviewing irrigation policy and rice cultivation since the early 19th century and attempts to link the pre-colonial, colonial and post-colonial experiences.

INTRODUCTION

The modern state of Malaysia was formed in 1963 by the union of the Federation of Malaya which achieved independence from Britain in August 1957, the island of Singapore which had been given internal self-government by Britain in 1958 and the territories of North Borneo (Sabah) and Sarawak, which had been British crown colonies since July 1946. On 9 August 1965, Singapore left the new Federation and became a separate independent state. Present-day Malaysia therefore comprises two territories separated by the South China Sea, West Malaysia (Peninsular Malaysia) and East Malaysia (Sabah and Sarawak).

West Malaysia forms part of the southern projection from the Asian mainland, with Thailand immediately to its north and the island of Singapore to the south. It has an area of 131,794 square kilometres and

---

[1] See for example, Ishak Shari and J.K. Sundaram, "Malaysia's Green Revolution in Rice Farming: Capital Accumulation and Technological Change in a Peasant Society", in Geoffrey B. Hamsworth, ed., *Village-Level Modernization in Southeast Asia: The Political Economy of Rice and Water*, Vancouver: University of British Columbia Press, 1982, pp. 225-54; James Fletcher, "Rice and Padi Market Management in West Malaysia, 1957-1986", *Journal of Developing Areas*, Vol. 23, No. 3, 1989, pp. 363-84; Richard Vokes, "State Marketing in a Private Enterprise Economy: The Padi and Rice Market of West Malaysia, 1966-75", Unpublished Ph.D. dissertation, University of Hull, 1978.

consists of eleven states: Perlis, Kedah, Penang, Perak, Selangor (with the Federal Capital Territory of Kuala Lumpur), Melaka, Johor, Negeri Sembilan, Pahang, Trengganu and Kelantan.

Physiography

The country consists of steep hill and mountain ranges, rolling to undulating land and coastal and riverine flood plains. The hill and mountain ranges cover about one-third of the surface of the Peninsula and run more or less parallel to the long axis of the country. The rolling to undulating land is found generally on the seaward flanks and the intervening areas between the mountain ranges. Although not very extensive, coastal plains and alluvial terraces are found from 15 to 65 kilometres inland from the coast with levels rising to 75 metres above mean sea level. The riverine flood plains are found as narrow belts of alluvium gently sloping away from the major rivers. Towards the coast they merge with the marine alluvium of the coastal plains.

Climate and Rainfall

Malaysia lies near the equator between latitude 1° and 7° North and longitude 100° and 119° East. The country is subject to maritime influence and the interplay of wind systems which originate in the Indian Ocean and the South China Sea. The year is commonly divided into the south-west and north-east monsoon seasons. The climate of Malaysia is hot, wet equatorial. The important features of the climate are the continuous warm temperatures and the seasonal distribution of rainfall. Mean daily temperatures range from 21°C to 32°C in the lowlands throughout the year. Cooler temperatures prevail at the higher altitudes. Variation in rainfall distribution is the most significant environmental variable. Generally most if not all parts of Malaysia experience moisture deficits during one or more periods of the year. Conversely excessive rainfall could occur and this may physically restrict agricultural activities.

There is considerable variation in the average for annual and monthly distribution of rainfall by location. The average annual rainfall ranges from 1,500 millimetres to 4,000 millimetres with the States of Sabah and Sarawak having about 20 per cent to 40 per cent more rainfall than Peninsular Malaysia.

Irrigation, Drainage and Rice Cultivation in West Malaysia

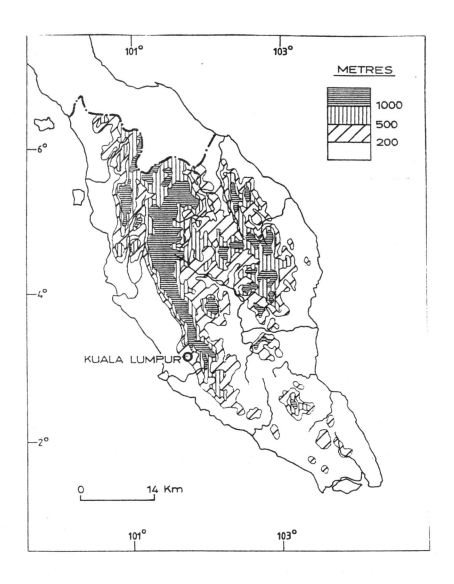

Map 1: Relief Map of West Malaysia

ECONOMIC AND POLITICAL FRAMEWORKS

The Malay Peninsula in the 1820s was, according to James Low, divided into two parts - agricultural and maritime. In the agricultural part, which included the states of Kedah, Perak, Patani and Trengganu, there was a considerable fixed agricultural population. In the maritime part, trade appeared to dominate. Unfortunately, Low did not identify the states with a more trading "bent".[2] In fact, nineteenth century Malay society was organised around agricultural production - swidden and sedentary. Swidden cultivation (*huma* or *ladang*/dry padi cultivation) involved periodic forest clearing for one or more seasons of cultivation after which the land was abandoned to revert to forest. Sedentary agriculture (*sawah* or *bendang*/wet padi cultivation) primarily involved wet-rice cultivation and the production of other food crops on adjacent land. Permanently settled communities were therefore only to be found cultivating *sawah* and were located on the coastal lowlands or coast areas of the Malay Peninsula. The larger and better endowed deltas of Kedah and Kelantan had a long history of permanent peasant settlements based on *sawah* cultivation.

In the nineteenth century, there was little specialisation in the Malay peasant economy and the typical Malay village produced a great deal of what it consumed. The peasant's life revolved around the river, the forest and a small area of land near his (her) dwelling. The land-holding was small - averaging two acres in some states. The basic economic unit was the family, engaged in mainly subsistence agriculture, in which rice and other local foodstuffs were essential items. A certain amount of produce was exchanged for such necessary goods as salt, ironware and textiles. Every member of the family, from the small child to the elderly grandparent, laboured to contribute to the family's wealth and periodic co-operative activities were often organised on a larger, sometimes village basis, for specific purposes. Land clearance and infrastructural construction, for example, tended to be communally organised, often in the form of *kerah* (forced labour or corvee). There was a general acceptance of co-operation with one's neighbours, and the value system of Malay villages stressed "co-operation, *gotong-royong* (co-operation), *usaha* (labour), and conformity".[3]

---

[2] James Low, 1836, *Dissertation on the Soil and Agriculture of the British Settlement of Penang or Prince of Wales Island in the Straits of Malacca*, Singapore: Singapore Free Press, pp. 82, 91 [Reprinted, Kuala Lumpur: Oxford University Press, 1972].

[3] Syed Hussein Alatas, "The Grading of Occupational Prestige Amongst the Malaysians of Malaysia", *Journal of the Malayan*

The basic agricultural activity was padi cultivation. The Malays grew both dry and wet padi. In *ladang* (dry padi) cultivation, the men felled the trees while the women cleared the brushwood and weeds. Sowing was a co-operative effort between the genders as was harvesting which also involved the help of neighbours. The technology used was simple. Under *sawah* (wet padi) cultivation the division of labour became more defined due to the introduction of the plough and the sickle. Tasks such as transplanting seedlings, weeding, reaping and the preparation, cleaning and drying of padi were done by women. The men prepared the fields for planting and also carried out the sowing, threshing and transporting of rice.[4]

Not all the states produced a surplus. Only the more settled and better endowed valleys were able to export rice to neighbouring areas. Kedah, for example, supplied rice to Penang and its neighbouring states. At one time, it exported 2,000 *coyan*[5] of rice to Penang, but after 1820 rice exports fell to about 100 *coyan* of padi and 50 *coyan* of rice.[6] When Frank Swettenham visited the state in 1889, he remarked that "the padi fields are of greater extent than any that I have seen elsewhere in the Peninsula. The whole country up to Perlis for some distance from the coast is one vast padi plain ..."[7] The importance of rice in the Kedah economy, apart from its export trade, can also be seen in the Kedah Laws. Out of four codes, two directly reflected the agricultural bias of the State. In fact, the second code (the Laws of Dato Sri Paduka Tuan - dated 1667) deals with various aspects of

---

Branch, Royal Asiatic Society, Vol. 41, No. 1, 1968, p. 153. See also J.M. Gullick, *Malay Society in the Late Nineteenth Century: The Beginnings of Change* (Singapore: Oxford University Press, 1987), Chs. 5-7.

4   R.D. Hill, *Rice in Malaya: A Study in Historical Geography* (Kuala Lumpur: Oxford University Press, 1977), Chs. 4-6.

5   (*Koyan*) measure of weight or capacity equivalent to 800 *gantang* or 40 *pikul* (1 *pikul* = 100 *kati* = 133 1/3 lbs = 61.7613 kg.)

6   James Low, *The British Settlement of Penang*, pp. 155-156.

7   See, for example, Sharom Ahmat, *Tradition and Change in a Malay State: A Study of the Economic and Political Development of Kedah 1878-1923*, Monograph No. 12 (Kuala Lumpur: Malaysian Branch of the Royal Asiatic Society, 1984), p. 17.

agriculture, especially with the cultivation of wet rice.[8] Additionally, the emphasis on canal construction reflects this importance.

The Kelantan Plain was another important rice cultivation area. In 1837, Abdullah bin Abdul Kadir, the writer of *Kesah Pelayaran Abdullah*, notes that a splendid rice crop was obtained annually from the broad coastal plain of Kelantan. This plain was low-lying and flat with fertile alluvial loam. Kelantan was not only self-supporting, it also exported rice to Singapore. The peasant cultivator also reared cattle which were exported on the hoof to Singapore and Thailand.[9] Other states which produced a surplus in the early nineteenth century were Perak and Trengganu. However, by the end of the century, production had declined. In Perak, demand exceeded supply, while in Trengganu there was land shortage and a rapid build-up of population in the Trengganu valley.[10]

Up until the middle of the nineteenth century, Selangor had an abundant supply of rice and did not need to import it. This was primarily due to the ruler's efforts (Sultan Muhammad 1826-57) who made it obligatory for his subjects to plant rice.[11] However, the Civil War of 1866-73 resulted in devastation of fields and great loss of lives which led to declining production.

The majority of the rice cultivators worked within a self-sufficiency framework. To supplement their diet, they reared poultry and grew fruits and other crops in their orchards. These crops included banana, sugar cane, tapioca, coconut and areca nut. In order to obtain basic necessities such as cloth, salt and tobacco, the peasants collected jungle produce like cane, bamboo and damar which they sold at the market place or fair. The forest also supplied them with fuel, fencing and roofing materials and leaves for basket- or mat-weaving.

Why did the Malays only engage in subsistence cultivation? Many reasons have been put forward by observers and scholars to explain this phenomenon. One explanation was the shortage or limited labour available to the peasant. As noted previously, the basic unit of production

---

8   See Richard Winstedt, "Kedah Laws", *Journal of the Malayan Branch, Royal Asiatic Society*, Vol. 6, Pt. 2, June 1928: 1-44.

9   P.L. Burns and C.D. Cowan, *Swettenham's Malayan Journals 1874-1876* (reprint, Kuala Lumpur: Oxford University Press, 1975), p. 268.

10   R.D. Hill, *Rice in Malaysia*, p. 69.

was the family. Consequently, most cultivators could not cope with an area beyond their production capability. Another explanation was the absence of draught animals in the country to do the heavy work of ploughing. Bullocks only became important with the introduction of oxen transport in conjunction with the rapid expansion of tin-mining in the last quarter of the nineteenth century. A third reason was the onerous burden of *kerah* or corvée labour demanded by the ruling class. In the mining areas, for example, the peasants had to work the mines at prescribed hours. Thus the peasants were never certain about their obligatory tasks and were therefore discouraged from expanding cultivation. Fourthly, irrigation facilities were simple and restricted to a few areas. Finally, although in most states a peasant was required to deliver a tenth of his produce to his chief as fulfilment of his traditional obligations, the uncertainty remained that a larger harvest could encourage the chief to be more rapacious and demand more.[12]

Political Changes

The British acquired Penang, Melaka and Singapore in the Straits of Melaka between 1786 and 1824. Commerce was the main activity in these settlements and there developed a merchant class comprising mainly British and Chinese British subjects. The three settlements were governed as a part of (British) India until 1866 when they were transferred to the Colonial Office and constituted as a Crown Colony in the following year. In the mid-nineteenth century relations between the Straits Settlements and the neighbouring Malay states were limited to commercial ties. Traditional British policy was opposed to any active interference either in or on behalf of, a Malay state. In practice, however, a fair amount of such intervention did take place. Furthermore, large numbers of Chinese had migrated to the Western Malay states to work in the tin mines or start agricultural enterprises. The merchants in the Straits Settlements believed the Malay states to be rich in mineral resources and also viewed them as having agricultural potential. Consequently, they urged the extension of British rule to these states.

---

11 Khoo Kay Kim, *The Western Malay States, 1850-1873: The Effects of Communal Development on Malay Politics* (Kuala Lumpur: Oxford University Press, 1972), p. 41.

12 See, for example, Sharom Ahmat, *Tradition and Change*, p. 21. See also Amarjit Kaur, "The Malay Peninsula in the Nineteenth Century - An Economic Survey", *Sarjana*, Vol. 4, June 1984: 69-86.

In the second half of the nineteenth century political conditions in several of the Malay states became unsettled to the point of chaos as a result of rivalries among Malay chiefs over the possession of important tin fields. Related disputes between antagonistic Chinese secret societies together with the growing importance of the western flank of the peninsula subsequent to the opening of the Suez Canal, led to British intervention in Perak, Selangor and Sungei Ujong in 1874. The role of the Straits merchants and entrepreneurs in this expansionist movement can hardly be overemphasised. British Residents were in due course appointed to each of the three territories. Subsequently, British protection was extended to the other territories of the Negri Sembilan and to the state of Pahang. The extension of British control paved the way for the large-scale development of the tin resources of the interior, particularly by Chinese entrepreneurs. Both the metropolitan and local government aimed at creating a favourable environment for private enterprise, whether European- or Asian-owned though the practical effects of the mining legislation tended to favour Western methods of operation. After 1912 Europeans replaced Chinese as the main large-scale miners.

In 1896 Perak, Selangor, Negri Sembilan and Pahang were joined in a federation called the Federated Malay States (FMS). A federal capital was established at Kuala Lumpur in Selangor and federal departments were soon formed. A Resident-General was appointed as the head of the FMS. Each of the Residents was responsible to him and he in turn was responsible to the High Commissioner for the Malay States who was the Governor of the Straits Settlements. A highly centralised administration was created to promote progress in the form of capitalist enterprise. After 1900, as a result of various economic forces working both inside and outside Malaya, Europeans also replaced Chinese as the main large-scale agriculturalists in the FMS.

In the north of the Peninsula, the states of Kelantan, Trengganu, Kedah and Perlis exchanged Siamese (Thai) for British suzerainty in 1909. In the south, Johor accepted British suzerainty in 1914. These five states accepted British Advisers (in Johor, a General Adviser) responsible to the Governor of the Straits Settlements. Uniformity in legislation and administration between each of the unfederated states and with the FMS and the Straits Settlements was obtained by the enactment of similar statutes and by the use of British administrative officers seconded from the FMS or the Straits Settlements.

The economic development of the west coast Malay states, particularly Perak and Selangor, progressed rapidly after the establishment of British protection. Initially, foreign economic activity was limited to the development of the most profitable tin resources in the most profitable locations sketched out by the location of tin deposits, rivers and the proximity of the Western states to the Straits ports. Since immigration was unrestricted, large numbers of Chinese entered the states to work the tin mines. Thus the western half of the peninsula experienced development which was carried out by foreign enterprise with the support of the government.

At the turn of the century, rubber became the most profitable plantation crop in the Malay Peninsula. Its distribution and spread was closely related to the extension of transport facilities and hence economic growth was again concentrated in the West coast states. This resulted in the creation of export-oriented enclaves and the associated infrastructure concentrated in the western half, leaving the eastern states outside the mainstream of capitalist development.

Government revenue was derived mainly from taxes on tin and rubber exports and those enterprises devoted to satisfying the needs of the Chinese mining population (opium, gambling). The Residents levied tolls on roads, leased out opium, gambling and other franchises, especially to Chinese entrepreneurs, and also taxed imports of rice, tobacco, spirits and salt. Their principal expenditures were on public buildings, roads, bridges, salaries for the administrative staff and political pensions to the Malay chiefs. The allocation for the improvement and creation of transport facilities was unusually high owing to the absence of any existing well-developed land transport system. Projects that did not bring in revenue were not given priority. Consequently, although the British viewed the Malay peasantry as providers of food for themselves and the labouring immigrant races, relatively little technical assistance was given in the form of irrigation facilities for the padi sector. This resulted in two major "rice crises" which in turn set the stage for planned rural development and the provision of better irrigation facilities under the aegis of the Drainage and Irrigation Department.

Briefly, colonial policy resulted in unequal development between the plantation and mining sectors on the one hand and the padi sector on the other. Rice production kept pace, though barely, with population increase and rice consumption. Thus, while major padi schemes were opened up, Malaya was unable to achieve self-sufficiency in rice.

Irrigation, Drainage and Rice Cultivation in West Malaysia

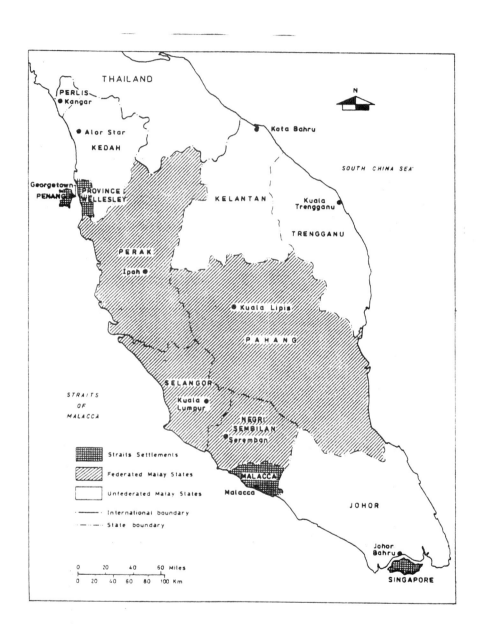

Map 2: Historical Map of West Malaysia

RICE POLICY, IRRIGATION AND RICE CULTIVATION

The extension of British political hegemony led to major economic changes. Vast areas of land were opened up for mining and agriculture. This was accompanied by an influx of immigrants from China and South India who provided the labour resources. In this set-up, therefore, there was a large and growing number of people engaged in non-agricultural pursuits who had to be fed. The British administration envisaged that the Malay peasantry would cater to the food requirements of the new labour. This policy would also ensure some sort of a balance between cash and food crops and at the same time meet the objective of rural welfare and stability. Consequently, by the 20th century, about 15 per cent of the total cultivated area was planted with padi, the second most widespread crop after rubber. Padi cultivation was mainly concentrated in the northern states of Perlis, Kedah, Kelantan, Trengganu and Province Wellesley and padi was an important crop in the other states. It was associated with the Malays and relegated to a secondary and subsistence position *vis-à-vis* the internationalised and predominantly expatriate export sector.

Irrigation and Rice Cultivation

Irrigation in Malaysia, like irrigation elsewhere in Southeast Asia, has been tied to rice production. In fact, the *raison d'etre* of an irrigation policy, whether traditional or colonial, was a concern with rice production. Prior to an official irrigation policy, irrigation works were poorly constructed and localised in nature. Sophisticated techniques of water management were only started in West Malaysia in the late nineteenth century.

As noted earlier, shifting cultivation predominated in most of the Malay states and it was only in the coastal plains of Kedah, Perlis and Perak on the west coast and Kelantan on the east coast that *sawah* or wet rice cultivation was practised. Here there was a settled agricultural population and more intensive farming techniques were used. On the whole, peasants were not inclined to invest voluntarily in irrigation works of a permanent nature. Rulers in turn could not depend sufficiently on their subjects to justify the construction of water control projects. Consequently, pre-modern water control systems were often crude and of poor quality. They varied regionally, depending on the physical environment.

Indigenous water control was based on the principle of conserving or raising the water level in a naturally swampy (*paya*) area with low embankments to the more evolved diversionary weirs (*ampang*) and water-wheels (*kincir*). Drainage channels, from ditches to full-scale canals (*sungai korok*) were regular features of the padi landscape.

The most widely used early method of water control was the brushwood dam or weir. It was built in a stream above or adjacent to the padi field to be irrigated. These dams, which relied on gravity flow, either diverted the water directly on to the padi fields or fed it into short distributary canals (*taliair*). The dams consisted of stakes interwoven with brushwood and were ballasted with mud and boulders. They were basic and temporary and any serious flood could result in their destruction. In swampy or *paya* areas, earth bunds or embankments were constructed to retain stream water.

Yet another structure was the bamboo water-wheel or *kincir*. These were used either ancillary to, or independent of the brushwood dams to transfer stream water to the fields. The water-wheel was particularly characteristic of valleys in the western foothills that had been settled by mainly Sumatran immigrants. This structure was also vulnerable to flood damage but could be dismantled at the end of the irrigation season.

These early structures were useful in areas of low population density and serviceable where catchments were relatively underdeveloped and still forested. Their main fault lay in their tendency to contribute to the deterioration of stream courses in conjunction with the rapid opening up of the country.

In the northwestern plains of Kedah, where sedentary agriculture was practised, water control structures were of a more permanent nature. In the early nineteenth century, the Sultan of Kedah used forced labour to build a canal between the Merbok and Muda rivers at the time of the Siamese invasion in 1821. Subsequently, in the second half of the century, the Sultan invited Malay aristocratic entrepreneurs to open up land for colonisation in the swamp forests of coastal Kedah. These aristocratic entrepreneurs were attracted by the potential market for rice in the Straits Settlements. The most famous of these entrepreneurs was Wan Mohamed Saman. In 1885, the Sultan granted him a concession to make a 32.2 km long drainage canal through the swamps south of Alor Setar at his own expense in return for a strip of land twenty *relong*[13] deep on either side which could be sold to colonists at $3.00 per square *relong*, plus a charge of 50 cents per *relong* rent. The canal was aligned with the aid of brushwood bonfires and beacons and apparently built by paid Chinese labour and forced Malay labour under the corvee or *kerah* system.[14] It proved successful financially and both its builder and others were encouraged to construct further canals. Wan

---

[13] (*orlong*) measure of length or of area - 240 feet or 73 metres approximately in length or 1 1/3 acres or 0.6 ha approximately in area.

Mohamed Saman himself undertook an elaborate project to build a canal from Sungei Pendang at Tanah Merah, 19.2 km above Alor Setar, via the Sungei Kangkong to the sea. Unfortunately, this venture failed, due largely to imperfect levels. At the same time, other concessionaires built similar canals with the result that by the end of the nineteenth century, an extensive but disjointed canal network emerged on the Kedah coastal plain.

These various canals or *sungai korok* did not connect with each other because "no concessionaire-holder was prepared to cut his canal through the unremunerative twenty-*relong* strip belonging to another concessionaire-holder."[15] After 1898, the works reverted to or were purchased by the government. Later in 1909, when a British Adviser was stationed at Kedah, the new advisory administration inherited the basis for a more comprehensive development of water-control facilities.

Modern techniques of water control were introduced by the British soon after the establishment of British political control in the peninsula. The introduction of these techniques was linked to the opening up of the country to western mining, plantation and commercial interests. Three phases may be distinguished in colonial irrigation policy and these phases are closely related to the extension of British power and the expanding capitalist economy. During the first phase (1874-1908), irrigation policies were localised and piecemeal, consistent with the localised penetration of the western Malay states. The second phase (1909-1931) saw the diffusion of modern water control methods to the other states. This phase coincided with the expansion of the capitalist economy and its dominance over agriculture and mining. The third phase (1932-1957) began with the establishment of the Drainage and Irrigation Department in 1932 in the Federated Malay States and the evolution of a national rice policy. It was concomitant with the consolidation of British power and influence and the transformation of Malaya into an export-oriented economy.

The First Phase 1874-1908

The initial projects undertaken by the colonial authorities included provision of small irrigation works and water gates to facilitate peasant agriculture. Another measure was to bring about efficient cultivation by synchronising standard dates for the various stages of padi cultivation. However, the comparative advantage of export-oriented agriculture over rice cultivation resulted in a very limited expansion of rice production, despite increased consumption by the expanding labour force employed in

---

14   *Annual Report Kedah 1909*, p. 24.

the export industries. A major reason was that padi farming was unremunerative and imported rice from Burma, Thailand and French Indo-China was cheaper than locally-produced rice. Writing of Negri Sembilan in the 1890s, Gullick[16] quotes figures which suggest that in an average year, padi growing was unremunerative compared to wage labour and that even in times of scarcity, a day's wage would buy a *gantang*[17] of rice.

Hence, there was no comprehensive plan for improving the country's self-sufficiency in rice prior to 1890. Whatever irrigation work was carried out depended largely on the initiative of local officials. These officials, however strong their enthusiasm, lacked data on specific irrigation requirements and techniques for rice cultivation in the Malayan context. They could only adapt and improve existing techniques in established padi lands and extend them to new ones. At this level and scale of operation, it was possible to use the system of "*mukim*[18] work" whereby government might supply equipment and limited funds while the peasants under their headmen would be expected to provide labour. This procedure, while it did permit modest achievement, tended to perpetuate the faults of traditional systems and in small amounts here and there dissipated considerable sums to marginal effect. The weakness in the procedure was aptly summed up by C.W. Sneyd-Kennessley, Acting Resident Councillor, Melaka, in his Report for 1893:

> "It is sad to see large tracts of land which if scientifically drained and irrigated would yield good paid crops, lying waste. The records show that thousands of dollars have been spent on amateur drainage schemes, many of which have not only proved failures, but have made things worse. Dams have been constructed in the wrong places or else, when made have been abandoned to decay; drains and canals have been cut which proved useless, while useful ones have not been kept up."[19]

---

[15] *Annual Report Kedah 1909*, pp. 24-25.

[16] J.M. Gullick, "The Negri Sembilan Economy of the 1890s", *Journal of the Malaysian Branch, Royal Asiatic Society*, Vol. 24, No. 1, 1951: 38-55.

[17] *Gantang* is a volumetric measure for padi equal to approximately 5 1/3 pounds (1 lb. = 0.453592 kg).

[18] Parish, area served by a mosque: (later) sub-district.

[19] *Annual Report Straits Settlements 1893*, p. 115.

Nonetheless, progress was achieved, particularly in certain coastal areas of Perak. The work there was of a slightly different nature in that it included the opening up of new land for mainly immigrant settlers. Noel Denison's work serves as a useful example. Not only does it illustrate the considerable degree of latitude left for the initiative of local officers but it also suggests the amount that could be achieved by a man of outstanding ability and enterprise. However, it also reveals the suspicion with which government regarded any method for rural development which was not obviously nor immediately financially rewarding.

Noel Denison was appointed District Magistrate of Krian in 1877 and soon showed considerable enthusiasm for settling and developing the country. He made an early attempt to improve an abortive irrigation canal constructed prior to the establishment of British Administration by a local chief. This canal, which ran from the lower Kurau River to the coastal padi lands of Krian was widened and deepened by the Public Works Department but the work was not very satisfactory. Undeterred, he pressed on with his ideas for improving the the area and encouraging the entry of settlers, using *mukim* work to drain further areas and providing control gates whereby water could be poured onto the padi field. When he was promoted to the post of Superintendent, Lower Perak, in 1891 he introduced a system of land settlement to Sitiawan based on the provision of loans to Banjarese and Javanese immigrant settlers. The major inducement was a loan at the rate of £17.3 per hectare as felling and drainage proceeded, thus allowing settlers to maintain themselves in the difficult and initial pioneering phase. Sitiawan was virtually uninhabited in 1889; under Denison's scheme, its population rose to 3,300 in 1893, at a total cost of Government (that is, outstanding subsidy) of only £328.35. In 1899, suggestions for irrigation works on a large scale were made and the Public Works Department was entrusted to collect the necessary information for tapping the Kurau River. The scheme was shelved because the government felt that the cost was prohibitive.[20]

In the last decade of the nineteenth century, interest was rekindled with the publication of the *Report on the Rice Supply of the Colony and Native*

---

[20] D.E. Short and James C. Jackson, "The origins of an Irrigation Policy in Malaya: A Review of Development Prior to the Establishment of the Drainage and Irrigation Department", *Journal of the Malayan Branch, Royal Asiatic Society*, Vol. 44, No. 1, 1971: 84; *Annual Report Drainage and Irrigation Department*, 1939: 2.

States,[21] commissioned by the Straits Settlements Government in 1891. The Report was a landmark in the history of official interest in water control in Malaya.

The *Rice Supply Report* comprised a series of reports on the extent and nature of existing as well as potential rice lands, together with suggestions on how to improve them. The report grew out of a concern at the large expenditure on imported rice by both the Straits Settlements and the Protected Malay States. Selangor alone, for example, imported M$1.5 million worth of rice in 1890.[22]

The Report furthered the interests of land improvement and water control for rice cultivation. Firstly, it drew the attention of officials to the fact that the government was interested in the improvement of rice cultivation and might be persuaded to allocate limited funds accordingly. Secondly, it enabled local officers to locate and record areas of potential padi land. This happened in numerous instances, one of the most noteworthy being Kuala Selangor where areas were designated and promoted for rice cultivation. Thirdly, the Report gave direction to the erstwhile disjointed ideas and localised efforts of certain officers who at *mukim* and district level, had been able to mobilise considerable resources for development on their own initiative.[23] Finally, it encouraged the preparation of a major water control scheme - the Krian Project.

The Krian Project

Krian was the first of many large-scale irrigation schemes in Malaya. Its construction pioneered the techniques which formed the basis of subsequent irrigation works in the country.

---

21   Straits Settlements, *A Report on the Rice Supply of the Colony and Native States 1893* (Henceforth referred to as *Rice Supply Report*, printed in the Proceedings of the Straits Settlements Legislative Council, 1893).

22   *Rice Supply Report 1893*, p. C105.

23   For example, Denison submitted a complete scheme which included the construction of weirs, channels and regulating gates. He also laid down principles by which he would have liked the government to open up land. He wanted the government to offer free or assisted passages to immigrants, waive land assessments for the initial three years of settlement, offer guaranteed work during the first six months of settlement, drain the land, and where necessary, provide irrigation facilities. *Rice Supply Report 1893*, pp. C132, C209.

The Krian area consists of a southward extension of the coastal plains of Kedah and Province Wellesley. It has slight gradients and heavy alluvial soils - eminently suitable for padi cultivation. In 1874, the area was sparsely populated. When Noel Denison was appointed District Magistrate from 1877 to 1881, he found Krian mainly all jungle, with a population of about 7,650 (1879).[24] In order to develop the area, he improved drainage and communications and waived initial land assessment for both padi and sugar planters. This resulted in an influx of peasants from Penang, Province Wellesley, Kedah and Pattani and included Banjarese from Indonesia. By 1888, the total population of Krian had increased to 16,600 and in the following year, some 14,837.2 and 8,094 ha were planted with padi and sugar respectively.[25]

These developments meant that rice cultivation had been pushed into areas where it could not be maintained by traditional or piecemeal water control methods. There were successive crop failures and in 1889, only 3,035.3 of the 14,837.2 ha alienated for padi were cultivated.[26] The Acting Collector and Magistrate, C.W. Welman, spoke up for systematic water control in Krian. Welman's proposal sparked off a debate on the form and desirability of a scheme to irrigate the Krian padi lands. Initially, it was believed that some form of combined water supply and irrigation system might serve the whole area. The task of collecting information was entrusted to the Public Works Department and in 1890, the Deputy State Engineer submitted a scheme for tapping the Kurau River. The scheme was dropped, partly because of the silt-laden nature of the Lower Kurau and partly because it was felt that irrigation was not an urgent necessity. Subsequently, further investigations were made on the drainage potential of the District and in 1892, a complete scheme was submitted which included headworks with flood spill weir, weir sluices and regulating gates together with 26.6 km of main and 66.8 km of branch channels. This scheme involved the diversion of the combined flow of the Kurau and Merah rivers by a weir at their confluence at Bukit Berapit on the eastern side of the plain. No storage was required because the average combined flow of these streams was 400 cusecs and this, combined with rainfall, was sufficient for 12,950.4 ha of existing and 8,094 ha of potential rice land. The estimated cost of the scheme was

---

24 Noel Denison, "The Kurau District", *Journal of the Malayan Branch, Royal Asiatic Society*, Vol. 18, 1886, pp. 349-352; *Annual Report Perak 1890*, p. 17.

25 *Annual Report Krian 1889, Perak Government Gazette*, Vol. III, 1890, p. 223.

26 *Ibid.*

$285,000,[27] and the government decided that its cost was prohibitive and postponed its construction.

In 1893, the Straits Settlements enquiry into rice supplies was published. The District Officer of Krian pressed for funds to allow him to revive the practice of opening up agricultural drains. Subsequently, it was decided to seek the advice of an expert from India. This expert, Claude Vincent from the Indian Public Works Department, was commissioned to prepare a report on irrigation in Krian. He visited Krian in 1894 and presented his report which did not differ greatly from the previous scheme except for minor details and notes on costs and returns. He proposed that existing drains be incorporated into the distributary system and a supplementary water supply be distributed ultimately to 21,044.4 ha of padi land. Where water rates were concerned, it was thought that a charge of one dollar plus 60 cents per acre ($1.48 per hectare) per annum would bring in a net return of 15 per cent on a capital outlay of $300,000.[28] Vincent's scheme was approved and a vote for irrigation works entered in the Perak estimates for 1895. On Vincent's recommendation, R.G. O'Shaughnessy, also from India, was appointed engineer-in-charge.[29] O'Shaughnessy undertook detailed new surveys and in his report dated July 1897, estimated the cost of the works at $860,000.[30] The scheme was temporarily deferred. It was revived again in 1898 by two local engineers, Ron Anderson and A. Murray. Subsequently, in August 1899, a scheme estimated at $785,000 was finally approved and work commenced. Construction proceeded continuously from 1899-1906 but estimates were revised annually. At the official opening of the Krian Scheme on 8 August 1906, it was announced that the scheme had cost $1,600,000.[31] Krian then accounted for nearly half the total wet rice acreage in the Federated Malay States (FMS). It had been provided with an irrigation scheme, which in subsequent years of low rainfall permitted good harvests in Perak. According to J.B. Carruthers, the first Director of Agriculture in the FMS, the scheme resulted in an increase of 30 to 40 per cent in the padi crop of the district.[32] Although improvements and extension works were carried out from time to time, much of the original

---

27   *Annual Report Drainage and Irrigation Department 1939*, p. 2.
28   Claude Vincent, "Report on Proposed Krian Irrigation Scheme", No. 2 in Government of Perak, *Collected Reports on Krian 1892-1930*, Taiping, 1932, p. 17.
29   *Annual Report Perak*, 1897, p. 15.
30   *Annual Report Drainage and Irrigation Department 1939*, p. 2.
31   *Annual Report Drainage and Irrigation Department 1939*, p. 2.

work remains. The only serious failure was the collapse of a portion of the flood weir in 1931, which necessitated its replacement by new gates that were constructed on solid foundations at the hill-foots on either side of the earthen embankment which impounded the water in the storage reservoir.[33]

The construction of the Krian Irrigation Scheme led to the realisation that provision would have to be made for laws governing the management and maintenance of the area. The first Irrigation Areas Enactment was passed for the state of Perak in 1899. This Enactment gave the Resident the right to "declare any lands within the areas affected by an irrigation works wholly or in part carried out or sanctioned by the government to be an irrigation area". The Act also empowered him to levy a water rate on the proprietors of land lying within such an area.[34] The Enactment was also intended to discourage the alienation of land in such areas for purposes likely to conflict with the aims of irrigation - that is, the planting of commercial crops as opposed to rice.

Thus, during the first phase of water management schemes, only one major scheme was built. This was consistent with government policy of constructing only those projects that would bring in revenue, or that had economic potential.

The Second Phase 1909-1931

As noted earlier, during the second phase the important rice growing states of Kedah, Perlis and Kelantan were brought within the ambit of British rule. This political change, coupled with the rice crisis immediately after the First World War, prompted the government to act on two levels. Firstly, the government took on a more active role in the construction of irrigation works in the FMS and the northern Malay States. Previously, all irrigation works were carried out by the Public Works Department and there was no specific department for irrigation. In 1913, an Irrigation Branch of the Public Works Department was formed. This branch was authorised to conduct investigations into the construction of irrigation works, the actual construction work being done by the district engineer. Subsequently, in 1920, the Irrigation Branch of the Public Works Department, FMS, was reorganised and known as the Hydraulic Branch of

---

32 *Annual Report Larut and Krian, 1907*, p. 3; *Annual Report Department of Agriculture, Federated Malay States, 1907*, p. 4.
33 *Annual Report Drainage and Irrigation Department 1939*, p. 3.

the Public Works Department. In 1921, the post of Chief Hydraulic Engineer was created. He was assisted by a staff of executive and assistant engineers. The duties of the Branch were extended to cover the investigation and collection of data in connection with the conservancy of rivers.[35]

The Northern Padi Plains

From the Kedah Peak to Perlis, practically the whole of the coastal plain is one vast rice field extending from the coast to the foothills. The soil is light alluvium, the best for growing padi, and the rainfall is more abundant during the rice growing period than elsewhere in Malaya, averaging 152.4-177.8 cm. A rudimentary system of drainage canals existed over large parts of the Kedah coastal plain and the state's role in supplying Penang with rice was long established. Under Siamese administration, the state owned canals had long been allowed to fall into disrepair. In 1907, with the arrival of the first qualified head of public works, the canals were earmarked for development with their banks to be adapted as potentially suitable for roads.[36] These states were relatively "poor" in comparison to the western Malay states. They did not have rich mineral deposits nor large tracts of land that could be alienated for commercial plantation agriculture. Consequently, funds were limited for construction of public projects like irrigation works.

The first change introduced by the British was at the administrative level, so that in terms of planning and the execution of projects, organisation was improved. Where irrigation works were concerned, only in Kedah, and to a limited extent in Perlis however, was substantial material progress achieved. The work was carried out by the Public Works Department. In 1912, the first government canal was cut through the swamp forest of southern Perlis using a combination of convict and *kampung* labour, the latter under the direction of the *penghulu*.[37] Perlis obtained the services of a Kedah engineer to advise and assist in this work so that the water control projects there were comparable on a small scale with those of Kedah.

---

[34] *Perak Government Gazette*, Vol. XII, No. 22, 1899, pp. 523-524 and No. 596, p. 624.

[35] *Annual Report Drainage and Irrigation Department 1939*, p. 7.

[36] See Zaharah Haji Mahmud, "Change in a Malay Sultanate: An Historical Geography of Kedah up to 1939", (unpublished M.A. thesis, University of Malaya, 1966), pp. 162-164.

[37] *Annual Report Perlis 1911-12*, pp. 36-37.

The first task in Kedah was to link up the canal laid before 1905 by indigenous entrepreneurs. This was completed by 1913. The next priority was to extend the canal system in order to expand the area of drained padi lands especially in the northern *gelam* swamps and in the southern districts pioneered by Wan Mat Saman, three decades earlier. Initially, progress was rapid because of the incentives offered during the First World War. In 1913, an engineer was posted to the Public Works Department with full-time responsibility for drainage and irrigation. By 1915, the network of canals maintained by the government had doubled.[38]

During the period of the First World War, there was an increased consciousness of the dangers of dependence on imported food. Between 1911 and 1916, for example, the FMS imported an average of 187,008 tonnes of rice annually or approximately 82 per cent of its annual rice consumption.[39] Furthermore, during the war, there was a shortage of shipping tonnage and therefore a need for restricting imports and exports. Tied to this was the urgency of increasing foodstuffs grown locally.

Between 1918-1920, the country faced a severe rice shortage and with decreasing rice exports to Malaya, rice prices soared to high levels. The government was forced to draw on its already depleted rice stocks and subsequently introduced a food control scheme whereby all rice stocks in the country were taken over by the government which assumed direct control over rice imports through a Padi Controller's Office. The rice was gradually released to wholesale dealers at fixed prices to ensure the most equitable distribution.[40]

The government also introduced legislation to boost local food production. The Malay Land Reservation Enactment of 1913, which aimed at preserving Malay land by restricting disposal rights, also placed cultivation restrictions on the use of reservation land. Its objective was to

---

[38] *Annual Report Kedah 1911-12*, p. 11; *Annual Report Kedah 1914-15*, p. 10.

[39] Memo from the Economic Botanist to the Director of Agriculture, Federated Malay States, 6 February 1918, encl. in High Commissioner's Office Files 387/1918.

[40] M$41 million was spent by the government to subsidise rice imports. *Report of the Rice Cultivation Committee*, 1931, Vol. 1, p. 14.

encourage the cultivation of padi as opposed to rubber.[41] The Rice Lands Enactment and the Coconut Palms Preservation Enactment introduced in 1917, were intended to prevent padi and coconut cultivation from being replaced or damaged by that of rubber. These were supplemented by a Food Production Enactment in 1918 which set aside land for the cultivation of food crops.

On the government's part, two main measures were taken to encourage rice production. One measure was aimed at the peasants. It included a scheme for the purchase and distribution of seeds, modifying the system of granting money advances to the peasants, paying bonuses to *penghulu* and chiefs who assisted in the food production campaign and sending Malay Sultans to exhort the peasantry to increase padi production.[42]

The other measure was the provision of technical assistance in the form of irrigation facilities. In 1921, the FMS government engaged the services of an expert on irrigation, a C.E. Dupuis. Dupuis' task was to report on projects for the improvement and extension of irrigation facilities. For the first time, extensive and detailed surveys were made of both existing and potential padi areas and water courses. Although his reports on irrigation and river conservancy projects were handicapped to some extent by lack of data, nevertheless, his report on Kedah produced some positive results. An Irrigation Branch of the Public Works Department was established in Kedah. Following this, between 1922 and 1931, controlled drainage schemes were carried out in the North Kedah area at a cost of M$1.5 million.[43] The general pattern was a process of consolidation and expansion of the water control network of Kedah and Perlis. The canals that were constructed included the Simpang Ampat-Arau Canal (Perlis), the Alor-Changeleh canal and the Sungai Korok Daun Canal (Kedah). The schemes centred on Dulang Kecil and the Kodiang/Sanglang rivers. There was therefore an extension from the older core areas in Kota Setar northwards along the coastal plain towards Perlis and southwards into Yen.[44] In 1918-20, the average area under padi in Kedah totalled 58,672.4

---

41 See Lim Teck Ghee, *Peasants and their Agricultural Economy* (Kuala Lumpur: Oxford University Press, 1977), pp. 103-138; Ahmad Nazri, *Malayu dan Tanah* (Petaling Jaya: Media Intelek, 1985).

42 Lim Teck Ghee, *Peasants and their Agricultural Economy*, pp. 122-123.

43 *Annual Report Drainage and Irrigation Department 1939*, p. 7.

44 *Annual Report Kedah 1927-28*, p. 28; *Annual Report Kedah 1929-30*, p. 18.

ha. In 1930-32, it had increased to 79,460.8 ha. Thus the provision of water control works produced a marked extension of the rice growing area.[45]

In the FMS, in terms of technical progress, aside from Krian, there were local projects only, depending on the ability of rural administrators to recognise the possibilities of and secure funds for small-scale drainage and irrigation schemes. In Pahang, for example, two water control schemes were prepared during this period, the Dong (64.8 ha) and Pulau Tawar (210 ha) schemes. In Perak, the Bota area on the Perak river came under review. Consequently, between 1914 and 1919, the area under padi in the FMS rose from 50,182.8 to 60,705 ha.[46] Reporting on both Kedah and the FMS, Dupuis had stated that there was a critical shortage of data upon which to base schemes. He therefore emphasised the important role of the Irrigation Branch as an agency for data collection and scheme preparation rather than execution.

In the 1920s, following Dupuis' report, a few schemes were started in the FMS. These included the Kenas, Pulau Tawar and Sri Menanti schemes. The clearing of the Krian reservoir of floating timber was also undertaken. Investigations also commenced on the possibility of developing certain areas of swamp in Perak, Selangor and Pahang for padi cultivation. In Perak, the scheme centred on a large area between the Perak River and Sitiawan (the Perak Tengah or Trans-Perak scheme). In Selangor, the Tanjung Karang swamp was surveyed (it had first been considered as a potential padi area in 1895). Here a dredger was required to cut drains before further investigations could be made but little progress was achieved until the formation of the Drainage and Irrigation Department (DID) in 1932. In Pahang, the Pulau Tawar scheme was completed in 1923, but it proved to be a failure. In Negri Sembilan, there was a long tradition of small-scale irrigation works and rural administrators had encouraged the construction or extension of these works. Dupuis suggested that government undertake the design and construction of small-scale works to replace some of the outmoded traditional stretches. In Rembau, the *waris* Fund, originally set up to promote the interests of clan inhabitants, was used to finance irrigation works. The first four of a series of dams were completed

---

[45] *Annual Report Perlis 1926-27*, p. 68.
[46] *Annual Report Department of Agriculture, Federated Malay States, 1913*, p. 13; *Annual Report Department of Agriculture, Federated Malay States, 1919*, p. 7.

by the Public Works Department in 1923 and these represent the earliest modern structures in Negri Sembilan.[47]

The Depression of 1929-32, which led to retrenchment in all government departments, also affected the Hydraulic Branch of the Public Works Department. During these years, only river and flood control works were considered a priority and no new drainage and irrigation works were undertaken.

Hence, during the period 1909 to 1931, drainage and irrigation works were only undertaken on a small scale in the FMS. In the north, the acreage under padi was greater than that under padi in the FMS. The FMS government continued to place greater emphasis on commercial plantation agriculture and mining and competing funds were channelled to these two sectors. Consequently, the rice problem intensified, and imports were consistently in excess of half a million tons annually, costing $50 million. About three-quarters of the rice consumed had to be imported.[48]

This period has been identified as the period of the "Great Malayan Rice Shortage". Essentially, this was due to the two-pronged colonial agricultural policy. Peasant farmers were encouraged to produce food for local consumption, while cash crop cultivation for export was intended to be a capitalist preserve. The main thrust of government efforts at promoting padi cultivation were negative. Investment in infrastructural development was minimal (compared to the investment in the export sector) and the government relied on legislation to impose cultivation restrictions on the peasantry to promote good production. The colonial authorities viewed peasant padi production as a necessity to reduce rice purchases from abroad and to reduce the foreign exchange required for these.

Thus, while rice consumption continued to increase over the 1920s, rice production did not expand significantly. Forced to respond to a continuing crisis situation, the colonial authorities established the Rice Cultivation Committee during the Depression in 1930. The Rice Cultivation Committee's task was to recommend means by which production of padi could be intensified and expanded. After prolonged deliberations and visits to the major padi districts of the Peninsula, the Committee presented its report in early 1931. In its report, the committee reviewed a wide range of measures - financial, agronomic, legislative and organisational. It

---

[47] Drainage and Irrigation Department, *Manual*.

concluded that "the important question involved in extending padi cultivation is the provision of better control in relation to water supply".[49] Thus the crux of the problem was seen to be the lack of any adequate government organisation for the preparation and execution of large-scale schemes planned in terms of a Malayan rather than a state or settlement basis. This led to the establishment of the Federal and Straits Settlements Drainage and Irrigation Department in 1932.

The Third Phase 1932-1957

The establishment of the Drainage and Irrigation Department (DID) in 1932 marked the beginning of a new era in irrigation policy in Malaya and the DID played a key role in consolidating existing padi production. The DID had an executive role in the Straits Settlements and FMS and only an advisory one in the Unfederated Malay States. In recommending its establishment, the Rice Cultivation Committee considered certain salient features of irrigation experience in the Malayan environment: the enhanced efficiency of large-scale schemes; the fact that state boundaries were not valid boundaries in either cultivation or irrigation; the existence of large areas of underdeveloped and potential padi land; the specialised technology required; and the need for intensive research into local irrigation problems. The DID's link with padi land development was stressed by the Committee.

The DID's principal functions and policy were:

(1) The improvement of irrigation and drainage facilities in existing padi areas with a view to increasing yields which would add to the country's rice production and at the same time promote the economic welfare of the peasantry.

(2) The development of new areas for smallholder settlement or for cultivation by mechanical means for increasing rice production on such a scale as would reduce the country's dependence on outside supplies.

---

[48] W.N. Sands, "Review of the Present Position of Rice Cultivation in Malaya", *Malayan Agricultural Journal*, Vol. XVIII, No. 3, 1930, pp. 131, 133.

[49] Federated Malay States, *Report of the Rice Cultivation Committee, 1931*, Vol. 1, No. 29, Summary and Recommendations, p. 24.

(3)　The construction of agricultural drainage works and the formation of Drainage Board Areas for crops other than padi, priority being given to smallholding areas. This would lead to increased food production and also raise the income levels of the Malays.

(4)　The maintenance and improvement within economic limits of the natural drainage arteries of the country with particular reference to agricultural interests (i.e. river conservancy schemes).[50]

Organisation of the DID

The establishment of the DID meant that there was now a specific department to foster the extension and improvement of rice cultivation.

In terms of organisation the officers of the Public Works Hydraulic Branch formed the nucleus of the DID, and ties with the Public Works Department, particularly in the early years, remained close, mainly for financial reasons. The first head of the DID was F.G. Finch, Acting Director of Public Works, and initial staff included seventeen engineers. Although established to overcome interstate boundaries, the DID staff reflected the political structure of the Peninsula. The FMS and Straits Settlements each had a separate section, coordinated by the Director's Office in Kuala Lumpur. The Department's role with regard to the Unfederated Malay States was advisory. Within the general framework, minor anomalies persisted until funds, personnel and work warranted their removal.

In Selangor, where there were few established works, the *status quo* with regard to the Public Works Department was retained, pending the commencement of the large-scale works at Tanjung Karang. Similarly, the vote for irrigation and drainage works remained a separate item of the Public Works Department budget. In the Straits Settlements the limited scope and potential of irrigation was reflected in the strength of the local Division, which included only one engineer, who was stationed at Melaka. Officers were not appointed in Penang and Province Wellesley until 1933. In 1936 the DID sections of the FMS and the Straits Settlements became more closely integrated and began to publish a joint annual report. The exclusion of the Unfederated Malay States from the executive sphere of the DID resulted in the northern and east Coast states (with the exception of Kedah) being handicapped by inexperience and lack of funds and having to

---

[50]　*Annual Report Drainage and Irrigation Department 1948*, p. 4.

rely upon the availability of visiting engineers in an advisory capacity. It was not until 1946 that executive sections of the DID were introduced to all the Unfederated Malay States, Kelantan and Trengganu sharing a single department.

During the period of the Japanese Occupation, the DID's organisation broke down. A proportion of the local pre-War staff continued to work on schemes in progress. However a shortage of materials for both construction and maintenance limited the amount of work achieved. No large new schemes were completed and existing works deteriorated through inadequate maintenance.

After the establishment of the British Military Administration, rehabilitation of irrigation and drainage works came within the sphere of the Royal Engineers. In April 1946, the DID was re-established with a return to civil government. Under the Malayan Union, administration of the DID was centralised with all states in the Peninsula coming under direct control of Kuala Lumpur. In 1948, however, the Union was reformed into the Federation of Malaya. Where drainage and irrigation was concerned, each state became responsible once more for maintenance of minor works, though large scale schemes remained under Federal financial and executive control. These schemes were to be handed over to state control on completion. The Director was, in a formal sense, once more an adviser to State DIDs but in practice work was more closely integrated than in the pre-War period.

In 1949, an important organisational change occurred when the need for long-term planning of irrigation and drainage works was acknowledged. A provision in ordinance schedule was made which allowed a five-year programme of work to be drawn up in accordance with the Draft Development Plan (1950-55) of the Federation. From 1952, the Annual Reports were published triennially to cater for the longer planning and development periods. On the administration side, recruitment of local citizens to senior staff positions within the DID was stepped up after 1950. The first Malayan Director was appointed in 1963, six years after Independence and the process of Malayanisation was completed in 1965. The close links between agriculture and drainage and irrigation were acknowledged when, in 1961, the DID was integrated within the Ministry of Agriculture and Cooperatives.

Achievements of the DID 1932-1957

The DID had a two-prong strategy to increase rice production in Malaya. One was the provision of irrigation and drainage facilities and the other the extension of padi lands. Other ways to increase productivity were limited. Double-cropping was not considered feasible and manuring experiments did not show encouraging results. However, seed selection was undertaken at the FMS agricultural experimental stations such as Serdang. In the 1930s the emphasis was on the extension of padi lands.

Prior to 1932 the colonial authorities had noted that there were several extensive tracts of potential padi land in the western Malay states. For example, irrigation schemes were considered for 2,023.5 ha in Kuala Selangor (estimated cost $120,000) and 44,517 ha in Bagan Datoh (estimated cost $2 million) but these had been shelved on financial grounds. The Rice Report of 1931 furthermore had provided an estimate of 240,000 to 4,000,000 ha of underdeveloped potential padi land. Included in this estimate were 100,000 ha in the Tanjung Karang area and a further 100,000 ha in Sitiawan which could be irrigated using water from the Sungai Perak. Various other areas were mentioned, including 20,000 ha of *hutan gelam* in Kedah and Perlis and 16,000 ha in the Endau Rompin area.[51]

The establishment of the DID provided the opportunity to carry out the development of such areas where they had not been utilised for alternate crops. The more accessible areas were the first ones to be developed. Some of the potential padi areas had already attracted settlers during the period of the Great Depression. The type of land sought for development had to meet various criteria. It had to be as flat as possible; have access to a supply of irrigation water and/or a suitable outlet to permit controlled drainage; and preferably have heavy soils relatively free from peat. In the west coast most of the areas which fulfilled these requirements had already been taken up by 1932. The areas that remained were those that had been designated earlier as potential padi areas and where surveys had already been undertaken. These were Sungai Manik and Changkat Jong in Perak, Tanjung Karang in Selangor, Paya Besar and Pahang Tua in Pahang and Padang Endau and Kahang in Johor.

Clearing and construction work began in the Sungai Manik area in 1933. Development took place in five well defined stages, covering 9,712.8 ha. This method was adopted in order to avoid construction work progressing at a rate faster than colonisation. Stage I was initially an independent unit with temporary irrigation facilities, and work was only started on Stage II

---

51 Federated Malay States, *Rice Report 1931*, Vol. 1, p. 22.

when the colonisation of the first stage appeared assured. The development of Stages II and III necessitated further temporary measures for irrigation and it was not until 1939 when the opening up of the first three stages was assured that a start was made on Stage IV. Construction then started on the main permanent headworks. By the end of 1941, the headworks and main canals to Stages I-IV had been completed, and the drainage and distribution system to Stage IV was well advanced. By 1941, 10,000 ha was under cultivation. The rivers that were involved were the Batang Padang, Batang Padang Mati (an outflow channel of the Batang Padang river), the Perak and the Batu.[52]

In Selangor, work began on the Panchang Bedana area for the draining of 6,000 ha. As at Sungai Manik, the area was partially flanked by smallholders established on the more accessible drained land along the coast. Investigations into the Tanjung Karang swamp continued until a scheme could be prepared in 1937, but the project was shelved due to lack of funding.

Between 1932 and 1940, construction began in Pahang on the Paya Besar Scheme involving 1200 ha. In 1938 another relatively small scheme of 1400 ha was started at Pahang Tua near Pekan. In both cases, difficulty was experienced in attracting colonists to take up the land.

The outbreak of the Second World War reawakened memories of rice shortages. Although earlier the poor economic conditions had led to a restriction of funds for construction, the war resulted in an accelerated diversion of funds to irrigation schemes in order to stave off rice shortages.

Funds were immediately provided for starting work at Tanjung Karang while surveys were rushed ahead at Changkat Jong in Lower Perak and Padang Endau and Kahang in Johor. Construction work proceeded during 1940-1941. At Sungai Manik, the rate of felling in Stage IV rose to 600 ha of forest per month in early 1940 so that by the time of the Japanese invasion, the whole of the 3,600 ha had been cleared and was awaiting irrigation. At Tanjung Karang, the Sawah Sempadan section and part of the Sungai Burung, totalling 4,600 ha were cleared during the same period. The remainder of the strip north to Panchang Bedena was partially intersected by drains and canals. Forest clearance proceeded at Changkat Jong with 900 ha by 1941 and 1,200 ha at Endau.[53]

---

52 *Annual Report Drainage and Irrigation Department 1946*, pp. 14-15.
53 Ibid., pp. 14-17, 38-39.

In essence, the DID's main contribution was to take over the farmers' own efforts at constructing and maintaining basic irrigation facilities. Additionally, the new schemes introduced also resulted in a shift in padi production centres in the FMS, from traditional Malay river valley settlements to the coastal plains peopled mainly by migrant Malays. This led to changes in the landscape of lowland Malaya with large tracts of forest being replaced by a rectilinear grid pattern of irrigation infrastructure and padi fields.

During the Japanese Occupation, a Japanese Director of Irrigation took over the reins. The Japanese, who were also concerned with self-sufficiency in rice, introduced double-cropping. They re-employed a large number of the subordinate staff, and each state functioned almost entirely as a separate entity with little central co-ordination. They continued with the work started in 1941, "more or less in accordance with the plans prepared by the Department", and "several concrete structures were completed". However the war and shortage of materials resulted in a deterioration in the maintenance of irrigation works. Thus the general trend was one of a slowing down of progress due to organisational failure and a shortage of materials and spare parts. At Tanjung Karang, a few access roads were constructed; at Sungai Manik, no progress was made and at Changkat Jong, the headworks were completed but remained inoperative.[54]

At the end of the Pacific War, Malaya was brought under a single central administration for the first time under British rule. Food supply was a major priority and the rehabilitation of the DID was carried out immediately. Colonial concern with padi production also led to the formation of a Rice Production Committee in the 1950s "to consider ways and means whereby the acreage planted under padi in the Federation and the yield per acre ... [could] be materially increased".[55] Although the Committee's 1953 Report emphasised double-cropping as a way to increase padi production, in the subsequent final Report of 1956, drainage and irrigation facilities were viewed as crucial for efforts to promote double-cropping.

Rehabilitation of the existing schemes was carried out systematically. Between 1946 and 1952, the Tanjung Karang scheme covered the rest of the Sabak Bernam-Kuala Selangor coastal clay belt, reaching its target of

---

[54] *Ibid.*, pp. 22-25.
[55] Quoted in R.H. Goldman, "Staple Food Self-Sufficiency and the Distributive Impact of Malaysian Rice Policy", *Food Research Institute Studies*, Vol. 14, No. 3, 1975: 251-93.

20,000 ha of new padi land. In Sungai Manik, the construction of canals and distributaries in Stage IV were completed by 1949. The proposed Stage V, which had reached final preparation, was postponed while borings for tin ore were sunk in the area. Although initial results were not fruitful, the irrigation works were not resumed. One of the reasons was the poor response to colonisation in Stage IV. At Changkat Jong, the works were completed in 1960 involving 2,000 ha.

Since 1942, only one large new scheme had been started i.e., the Seberang Prai (Trans-Perak) Scheme. Interest was first shown in the swampy area between the Perak River and the Dindings in 1876 and thereafter its agricultural potential was considered from time to time. In 1931, the Rice Report suggested that investigations should continue but priority for development was given to the more accessible areas such as Sungai Manik and Kuala Selangor. At that time, the estimate was 40,470 ha of padi land although colonists had planted rubber in areas adjacent to the roads. In 1952, a consultant was appointed to complete the detailed feasibility surveys of the area. It was assumed from the surveys that 41,000 ha of potential padi land were available for development. By 1960, it was apparent that much of the suitable land was in tracts scattered throughout the area with intervening areas of deep peat. The scheme eventually adopted was for the opening up of 15,000 ha of padi land in Stages I, III and IV, the rest of the Trans-Perak area being planned as a drainage area for crops other than wet rice.

The early colonisation schemes were based on the premise that the Malay peasants preferred to continue padi farming. In fact, many of them were drawn to rubber and left the padi fields untended. Consequently, the British allowed non-Malays to occupy some of the new rice lands at Sekinchan (Tanjung Karang) and Changkat Jong. The irrigated land colonisation schemes developed since 1932 represent one of the most fundamental modifications of the landscape of Malaya. The Trans-Perak scheme amounts to approximately 49,373.4 ha or 16 per cent of the total padi area of Malaya. Of this area the vast majority has been created in areas previously covered by original or secondary forest, much of it in a water-logged condition.

The Northern Rice Lands after 1931

The Rice Cultivation Committee acknowledged that improvements in existing padi areas were more likely to bring about immediate results than the creation of new areas. In this regard, apart from Perak, the northern states of Kedah, Perlis and Kelantan offered the best opportunity for this type of improvements. In the northwest, there were three main areas located in Kedah and Perlis. These were the rice fields in the extreme north of the region in Perlis and to the east of Alor Setar in Padang Terap; a large area between Arau and Alor Setar; and the area south of Alor Setar towards Yen. In 1935-36, the Perlis Public Works Department undertook extensive surveys of the Arau-Alor Setar area, and in 1936 prepared a scheme for the drainage and irrigation of over 6,475.2 ha to be known as the South Perlis Irrigation Scheme.[56] The proposal was to dam at intervals the Simpang Ampat-Arau canal, a variation on the inundation principle. However, on the advice of the DID, FMS, this plan was dropped. Instead the head-waters of the Arau, Gial and Jerneh rivers were diverted by headworks and distributed to the padi lands of both north and south Perlis in a separate system on the advice of the DID, FMS.[57] Certain works were completed before 1942, which included the Arau headworks and some canals and subsidiary drains.

In Kedah, a scheme to serve 30,352.5 ha of padi land was projected. As in Perlis, the principle was to construct drains, intercept short west flowing streams (the Tanjung Pau and the Padang Terap) with headworks at their point of emergence from the Main Range foothills and divert their waters into canals serving the plain. Because of the size of the task, surveys and plans were not completed before the Japanese invasion.

In the post-war period, the Perlis and Kedah schemes were revised and integrated to provide drainage, supplementary irrigation and sea defence throughout the areas. Irrigation water was to be supplied to the southern portion of the Perlis section from the headworks in Kedah via the Sanglang canal.[58]

Although reported as irrigation schemes it was widely acknowledged that their main functions were to drain the padi areas and distribute evenly the available supply of irrigation water.[59] Work began in 1947 on the Kubang Pasu section of the plain and in 1948 on South Perlis. The latter area was completed in 1957 by which time a new scheme had been formulated to bring the remaining section of the coast plain for north Kedah and Perlis into an integrated system of water control. The DID's policy on new land development was also modified and now confined to areas "... adjacent to or

---

[56] *Annual Report Perlis*, 1936, p. 29.
[57] *Annual Report Drainage and Irrigation Department 1938*, p. 62.
[58] *Annual Report Drainage and Irrigation Department 1948*, p. 17.
[59] *Annual Report Drainage and Irrigation Department 1949*, p. 10.

Irrigation, Drainage and Rice Cultivation in West Malaysia

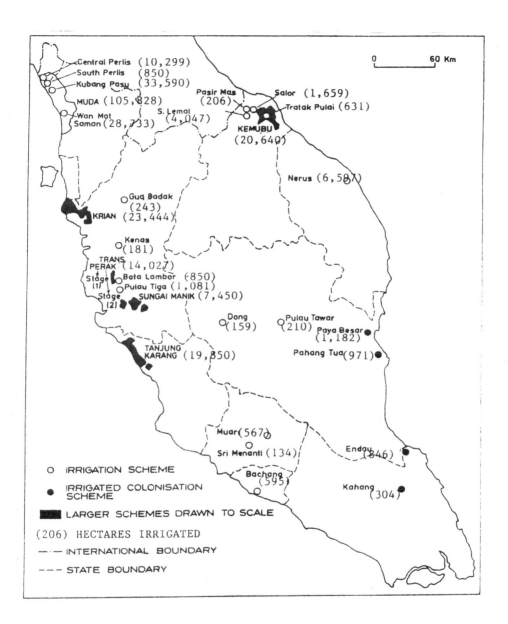

Map 3: Malaya - Irrigation Schemes 1931-57

surrounded by existing padi areas ...".[60] With the completion of Kubang Pasu in 1961 and Central Perlis in 1964, the whole of the north-west padi plain north of Alor Setar was served by a comprehensive system of water control schemes which, by regulating natural run-off within the area, supplemented by a limited irrigation supply, gave relatively secure conditions for single crop padi cultivation over approximately 58,681.5 ha. By 1964, virtually the whole of the north-western coastal plain was served by some form of large-scale water control works. The Wan Mat Saman Canal System of drainage and limited irrigation facilities serving the area south of the Kedah River, which had been earmarked for reconstruction in 1946, underwent progressive improvements from 1959 onwards. This took the form of improved outfall channels, colonisation of the Sala River and new coastal drainage control. The whole region, containing some 105,222 ha of padi land has since 1964 had the basic drainage infrastructure for the superimposition of a massive new project, the Muda Irrigation Scheme, resulting from the new approach to improved rice production - double-cropping (see below).

In Kelantan, the only established forms of water control were brushwood dams serving small areas. This was because a sizeable portion of the padi planted was *tugalan* or dry padi. The Rice Report identified Kelantan with Kedah in terms of potential land. In 1930, the Colonial Development Fund made a grant to the Kelantan Government for the investigation of a large-scale scheme for irrigating the Kelantan plain from a mainstream barrage or tank-storage in the foothills of the Kelantan River catchment. Because of the severely depressed economic condition of the state, it was decided to use a proportion of the funds directly for small-scale works. By 1940, there was, apart from small-scale improvement works, only one gazetted irrigation area in Kelantan of 607 ha at Tratak Pulai.[61] After the war, DID surveys showed that the Kelantan plain was not suitable for simple gravity schemes of drainage and supplementary irrigation. Therefore a new irrigation structure was required - that which involved the application of mechanical pumping.

Mechanical pump irrigation schemes were not new. The first large-scale pump scheme was developed between 1932 and 1935 in the Bota and Lambor *mukims* of Perak, south of Parit on the Perak River, following the 1931 Rice Report. The problem was to irrigate the numerous tracts of padi land,

---

[60] *Triennial Report Drainage and Irrigation Department, Malaysia, 1961-63*, p. 5.

[61] *Annual Report Drainage and Irrigation Department 1938*, p. 63.

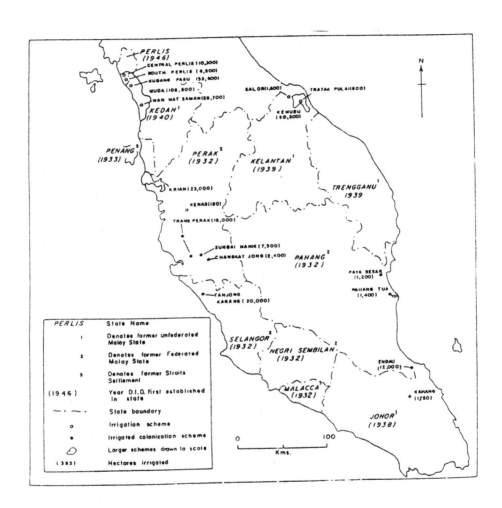

Map 4: Malaya Water Control Schemes 1931-1957

varying in size from 8 to 162 ha (1,012 ha in all) extending over 17.7 km, in the absence of any immediate possibility of a river barrage project.

The Bota scheme involved a main canal running south from a diesel pumphouse near Kampung Bota Kanan, linking the padi areas along a route parallel to the river for about 19.3 km. In 1935, a second pumped scheme was commenced at Pulau Tiga, about 4.8 km to the south of Kampung Lambor. This was completed in 1937. Despite initial problems, these schemes resulted in increased yields.[62] Subsequently, the pumping structure was used in Pahang Tua as a well.

In the post-war period, pumping assumed more significance in connection with the adoption of double-cropping as the solution to Malaya's rice problem. The 1953 Rice Report emphasised both the scale and nature of the potential of pump schemes in permitting assured supplies for double-cropping. The Report had not only the riverine padi tracts of the Perak Valley in mind but also the padi plains of the north, especially Kelantan.[63]

In 1946-47, plans were prepared for a pilot scheme to introduce pumped irrigation in Kelantan. The site chosen was at Salor on the Kelantan River to command 2,023.5 ha to the Pasir Mas district.[64] This project, the first to employ electrically-powered pumps in Malaya, was completed in 1951 and resulted in yield increases of over 100 per cent in comparison with adjacent padi areas.[65]

Apart from irrigation works, the government also sought to promote the use of chemical fertilizers as a means of increasing padi yields, particularly in the east coast states. In 1952, a scheme was introduced to provide subsidised fertilizers in Kelantan and Trengganu. Nonetheless, there was a low record of usage, in part due to poor price incentives, difficulties of access to west coast markets and the conservatism of the peasantry.[66]

Briefly therefore, irrigation schemes during the colonial period were constructed to serve both existing and new areas and the technical

---

[62] *Annual Report Drainage and Irrigation Department 1939*, p. 26.
[63] Federation of Malaya, *Rice Report 1953*, pp. 60-61.
[64] *Annual Report Drainage and Irrigation Department Malayan Union 1947*, p. 30.
[65] *Annual Report Drainage and Irrigation Department FM 1951*, p. 21
[66] *Annual Report Federation of Malaya 1954*, p. 138.

requirements for such schemes went through several stages of development. The initial stage of irrigation development was to provide a system of controlled drainage in existing cultivated padi land. While the abundant and well distributed rainfall in the country normally provided sufficient water for plant growth if field ridges (*batas*) were constructed to retain it on the land, the difficulty encountered was to drain off excess water at the time of harvest. Controlled drainage systems provided some level of water control especially for the coastal areas where the slope of the land was normally flat. Along the Pahang river, many *paya* (swamps) were formed by the silting of the river and the building up of its banks to a level higher than the country inland. These *paya* were usually depressions lying between the foothills and the river banks. Because they had small catchments, water for irrigation was scarce. Consequently, an "inundation" system of water control was introduced for padi cultivation in these areas. Low earth embankments together with control gates and flood spillways were constructed across the valley and to inundate the *paya* and supply the necessary water. These controlled drainage and inundation systems, while providing some means of water control, were at best rudimentary. Gradually, such schemes were improved with the provision of "positive" irrigation facilities whereby water was diverted from a river into a system of canals commanding the padi lands. Up to the mid-1950s, new areas such as the Krian, Sungai Manik, Tanjung Karang and Kubang Pasu Schemes were reclaimed and provided with irrigation facilities for padi cultivation.

The colonial authorities promoted peasant padi production for a number of reasons. In contrast to shifting cultivation, wet padi cultivation offered the prospect of permanently settled peasant farmers. Additionally, a peasantry catering for local consumption needs would not threaten the export-oriented plantation sector. And finally, increased domestic production would lessen Malaya's dependence on imported rice and thus reduce the foreign exchange outflows required for rice purchases from abroad.

Independence to 1982

Colonial policy had been extractive, not developmental, and consequently funds allocated to drainage and irrigation works were small in comparison to funds allotted for the development of the plantation agriculture and mining sectors. Huge amounts were spent on railway and road construction to facilitate the expansion of the capitalist economy. Nonetheless, it cannot be denied that rural development did take place and that funds were expended on drainage and irrigation schemes especially during rice crisis periods. The organisational set-up and irrigation structures formed the basis for further development when Malaya achieved independence in 1957.

Although the main thrust of agricultural policy in the post-independence period has been directed at export-earning crops, rice farming has received sustained attention and assistance by the Malaysian government. Such assistance includes the development of irrigation facilities, fertilizer subsidies, extension work in promoting the use of high-yielding varieties and double-cropping, establishment of marketing and credit facilities and protection of domestic producers through implementation of a guaranteed minimum price for padi and control of rice imports.

On the organisational/administrative level, the formation of Malaysia in 1963 resulted in the establishment of state DIDs in Sabah and Sarawak and this increased the number of State Departments to 13, with the Federal Headquarters being responsible for the overall drainage and irrigation matters in the country. In 1970/71, severe floods occurred in many parts of West Malaysia and the situation was so serious that a national disaster had to be declared. Consequently, flood mitigation was made an additional responsibility of the DID from 1972 onwards.

Generally, up to about 1960 the majority of the irrigation schemes were mainly designed and constructed to stabilise the production of a single crop of padi in a year against periodic water shortage in the rainy season. The main common feature was that the schemes were designed to supplement the rainfall to the extent of providing a reliable irrigation water supply throughout the growing season. A total of 213,600 ha were equipped with these facilities.

The early 1960s marked a new stage in irrigation development. This was the beginning of a period of accelerated irrigation development as an integral part of the rural and socio-economic development programme in the country following Independence. The thrust of the irrigation expansion programme was directed towards the development of water resources and provision of major engineering works for the cultivation of a second crop of wet padi during the drier months of the year. To bring about double cropping in existing major padi growing areas, the construction of storage reservoirs and large pumping stations was undertaken. In rain-fed padi areas, systems of conveyance and distribution canals and drains together with diversion weirs were provided to replace existing rudimentary works. For areas where basic facilities were already provided for single cropping, the existing reticulation systems were improved or enlarged to cope with the higher capacity requirements for growing the second crop.

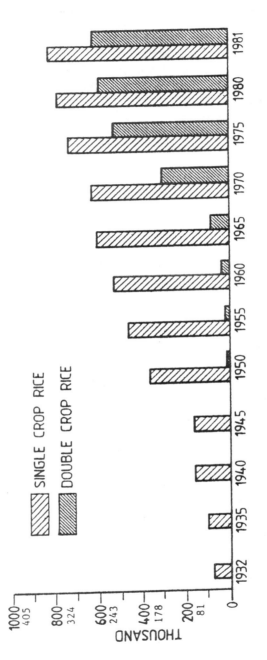

Figure 1 Area provided with Irrigation Facilities (in acres and hectares)

Note: The smaller figures denote hectares while the larger figures denote acres.

Up to 1970, the increase in padi land under double-cropping was achieved largely in areas with existing irrigation facilities. After 1970, two major irrigation schemes became fully operational and these brought a sharp increase in planted acreage. The first project was the Muda Irrigation Scheme in northwest Peninsular Malaysia which enabled double-cropping of about 105,827.6 ha of land. Since its completion, the Muda region has produced about 50 per cent of all locally produced rice marketed in Malaysia. Capital expenditure on this project amounted to M$228 million, of which approximately M$135 million (59 per cent of the total) consisted of a loan from the World Bank.[67] The second project was the M$40 million Kemubu Scheme which converted another 22,258.2 ha from single to double-cropping.

By 1975 all major rice growing areas were equipped with irrigation facilities to enable double-cropping to be carried out. These included the Muda, Krian, Sungai Manik, Sungai Muda, Tanjung Karang, Besut and Kemubu areas totalling some 190,000 ha. New areas such as the Trans Perak Stage I and parts of Stage IV were also reclaimed for double-cropping of padi.

Presently, agricultural development in the country emphasises an integrated approach involving considerations of engineering, social, agronomic, cultural and economic factors. Irrigation provides the basic favourable environment for the other components to be successfully introduced. The physical irrigation infrastructure provided includes the tertiary irrigation and drainage systems and farm roads. Since the Third Malaysian Plan, large irrigation projects such as Muda, Krian/Sungai Manik, Barat Laut Selangor and Kelantan Utara were planned and implemented based on this integrated approach.

The government's emphasis on irrigation development is reflected in the development funds allocated to the irrigation programme in the country's Five Year Plans, as indicated below. (See also Figure 2)

- First Malaya Plan (1955-1960) - M$ 25 million
- Second Malaya Plan (1961-1965) - M$ 69 million
- First Malaysia Plan (1966-1970) - M$100 million
- Second Malaysia Plan (1971-1975) - M$151 million
- Third Malaysia Plan (1976-1980) - M$314 million

---

[67] See Clive Bell *et al.*, *Project Evaluation in Regional Perspective: A study of an Irrigation Project in Northwest Malaysia* (Baltimore, Johns Hopkins University Press, for the World Bank, 1982).

Figure 2: Expenditure on Development Programmes

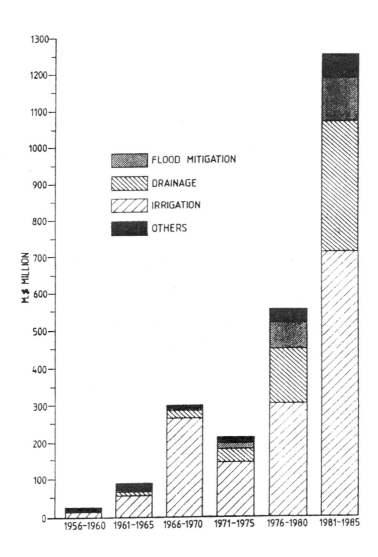

The bulk of the investment has been devoted to the provision of major engineering works such as storage reservoirs, pumping installations and main conveyance systems. Land and water are matters under State jurisdiction, and operation and maintenance costs of completed irrigation projects are met from the respective State funds. In accordance with the Irrigation Areas Ordinance 1958, the State collects irrigation water rates from the farmers from which the Operation and Maintenance costs are met. The impact of investment on irrigation works during the various five-year development plans can be gauged from the fact that the total rice production in Peninsular Malaysia increased from about 609,700 tonnes in 1960 to 929,100 tonnes in 1970 and to about 1,182,400 tonnes in 1974. Table 1 shows the increase in rice production during the period 1960 to 1974.

The large increase in total rice production was due to three main factors: an increase in yield per acre, an expansion of harvested area and the diffusion of double-cropping as shown in Table 2.

Table 1: Production, Consumption, and Import of Rice, Peninsular Malaysia

Year	Production ('000 tonnes)	Import ('000 tonnes)	Consumption ('000 tonnes)	Self-Sufficiency Rate (%)	Per Capita Consumption (kg)
1960	609.7	362.7	972.4	62.7	-
1967	706.2	290.5	960.2	70	117
1970	929.1	267.6	1,196.7	78	135
1972	1,017.9	99.1	1,117.1	91	120
1974	1,182.4	207.8	1,340.1	85	142

Source: Ministry of Agriculture and Rural Development, 1975, *Padi Statistics 1974*, Kuala Lumpur, p. 78; quoted from T. Ouchi *et al.*, 1977, *Farmers and Villages in West Malaysia*, Tokyo: University of Tokyo Press, p. 32.

Table 2: Increase of Total Production, Harvested Area and Yield of Padi

Year	Total Production* (mil. kg)	Harvested Area		Yield/Hectare	
		Main season ('000 ha)	Off-season ('000 ha)	Main season (kg)	Off-season (kg)
1952	570	304.7	1.6	720	685
1974	1,787.5	354.5	215.7	1,170	1,312.5
Increase	1,217.5	49.7	214	450	627.5
Rate of increase (%)	(213)	(16)	(13,225)	(63)	(92)

Source: Ministry of Agriculture and Rural Development, 1975, *Padi Statistics 1974*, Kuala Lumpur, p. 78: quoted from T. Ouchi et al., 1977, *Farmers and Villages in West Malaysia*, Tokyo: University of Tokyo Press, pp. 32 & 39.

Note: * Dry padi is excluded. Total production is not equal to harvested area.

The irrigation development programme has brought about improved agricultural productivity, thereby benefitting some 200,000 padi farmers living within the irrigation area. The irrigation investment is therefore effective in increasing the income of the padi farmers apart from the fact that it provides substantial indirect benefits.

THE ROLE OF DRAINAGE

Prior to the twentieth century, very little was done in respect of agricultural drainage except for the cultivation of padi. Land drainage for agricultural crops other than padi was carried out independently by plantation owners and small-holders cultivating coconut, rubber and other crops to serve their limited purpose and consequently there was little or no coordinated effort. The early reclamation works were constructed through communal effort whereby the land was protected from tidal inundation by an earth bund constructed along the coast. However, these structures were rudimentary and inadequate and coupled with the lack of proper maintenance, they failed in time. Destruction of inland cultivation thus occurred and large areas of coconut, rubber and small-holder crops were adversely affected.

In 1909 a Drainage Rate Enactment was passed in the Federated Malay States. The first drainage scheme was constituted in 1912 and known as the

Kapar Drainage area. Fourteen other areas along the Selangor coast were also declared drainage areas between 1914 and 1927. The original works were carried out by estates and these were added to from time to time or improved. In the coastal areas, the drainage works were necessary because of the entry of salt water to agricultural lands which killed off or interfered with coconut plantations and padi cultivation. The agricultural lands were originally opened up by construction of drains but without adequate protection from the sea, tidal waters swept the areas. Subsequent provision of coastal bunds and tidal control gates together with internal drainage channel systems allowed reclamation of these coastal agricultural lands and the opening up of new areas. Most of the early drainage schemes were carried out along the coastal areas of Selangor and Perak. In order to pay for the drainage works, the Government advanced the capital from time to time and levied a rate to cover 5 per cent interest and maintenance charges. Capital outlay at the end of 1909 was $103,900 and this had risen to $178,000 by 1923.[68]

In Perak, the Bagan Datuh Drainage area was the most important. Its history dates back to the early nineteen hundreds but its development continued only in a piecemeal fashion up to 1924 when it was gazetted a drainage area. When the Drainage Board took over the area they found it in poor condition and $367,000 was spent between 1925 and 1931 reconstructing bunds and additional water gates, the deepening of existing drains and construction of new drains.

In Selangor, the Sabak Bernam Peninsula area was of particular interest. Prior to 1924 settlers constructed sections of bund and dug drains and some small works were carried out by the Government to assist the cultivators. Between 1924 and 1927, further sections of bund and several water gates were constructed. The bunds were made by digging borrowpits adjacent to them and were of small dimension, having only one foot of freeboard above the Spring Tide High Water level. The water gates were of the screw-down wooden shutter type with insufficient length to prevent percolation and too high a cill causing severe downstream erosion. Abnormal gales coinciding with high tides in October 1931 caused severe damage to the area necessitating complete reconstruction which was carried out between 1933 and 1937.

In 1932 the newly formed Drainage and Irrigation Department took over responsibility for drainage. Between 1932 and the Japanese Occupation, extensive drainage and reclamation schemes were undertaken to improve

---

[68] *Annual Report Drainage and Irrigation Department 1939*, p. 10.

the *kampung* lands of small-holdings planted with coconuts, rubber, fruit etc. Most of the work during this period was carried out in the coastal areas of Selangor and Johor.

In Selangor, the main work comprised the protection of the coastal area from Sabak Bernam to Kuala Selangor and involved the construction of 96.5 km of bund, 18 tidal control gates and 144.8 km of main drains. The bulk of earthworks was carried out using dragline excavators. There were four main drainage areas in this zone, the Bernam Drainage Area in the north, and the Sungai Besar, Tanjung Karang and Ujung Permatang drainage areas to the south. Work commenced on the Bernam Peninsula area in 1933 and apart from a small part of the Tanjung Karang area, all works were completed by 1941 at a total cost of $680,000.

Work in Johor, due to the later establishment of the Department there, only commenced in 1939. Drainage schemes were prepared for two large areas, the Senggarang Drainage Area and the Sri Menanti Area. These areas included a large number of small-holdings mainly planted with coconuts. The schemes proposed the construction of coastal bunds and provision of tidal control gates for excluding sea water and improvement of internal drainage to alleviate flooding from inland waters. By the end of 1941, the main Senggarang drain had been constructed, the outlet gate completed and an excavator was engaged on construction of the Sembrong flood protective bund. At Sri Menanti, an excavator was employed on the construction of the coastal bund from the southern end and one out of five tidal control gates had been completed.

During the period of the Japanese Occupation (1942-45) drainage areas were largely neglected and drains became silted up and in some cases were almost entirely blocked. In the Bagan Datuh Drainage area, the perimeter bund was so neglected that it had to be largely reconstructed after the war.

Between 1946 and Independence in 1957, drainage programmes emphasised the maintenance of existing schemes and the continued construction of the Senggarang and Sri Menanti Drainage Areas which were completed in 1957. By this time, the total drained area in the Peninsula amounted to 140,000 ha.

After Independence further progress was made and the number of drainage schemes undertaken by the department increased rapidly. More than 60 per cent of these schemes were in the coastal regions (principally the west coast) and the main crops cultivated were coconut, rubber, areca nut, and coffee. A total of 35 drainage areas were maintained in 1963, the bulk of these being on the coastal areas of Selangor, Perak and Johor.[69] By 1973 the total drainage area amounted to 350,000 ha and in 1982 the figure was 476,000 ha.

Briefly, the drainage schemes provided protection against sea water intrusion and river floods. Facilities were also provided to remove excess runoff from local rainstorms while providing for some measure of ground water table control. Consequently, the schemes protected and prevented damage or destruction to commercial crops such as rubber, oil palm and coffee, which were widely cultivated on reclaimed lands. However, the provision of drainage facilities alone would not have been sufficient to improve production and concomitantly the incomes desired. To achieve this, agricultural support measures (such as replanting with high yielding variety and intercropping as in the case of small coconut holdings) were required to be implemented at the same time. The present development of large drainage areas proceeds along an integrated approach, involving various agencies. Projects planned and implemented like the West Johor, Nonok, North-West Selangor, and Bagan Datuh are notable examples of integrated projects currently being undertaken.

Furthermore, in recent years, with the successful introduction of cocoa intercropping in many coconut replanting areas where the soil was found suitable, the facilities in many drainage areas designed on past criteria became inadequate to meet the more stringent requirements of these intercrops. This called for drainage and intensification works of which a typical example is the North West Selangor Integrated Development Project where 44,517 ha of small-holder tree crop land were provided with more intensive drainage which permitted replanting of some 35,209 ha of coconut land with improved coconut varieties and cocoa.

Where financing is concerned the government investment on drainage improvements since independence has increased steadily and this is reflected in the increasing flow of funds under the Drainage Programme as summarised below:

- First Malaya Plan   (1955-1960) - M$ 11 million
- Second Malaya Plan   (1961-1965) - M$ 34 million
- First Malaysia Plan   (1966-1970) - M$ 90 million
- Second Malaysia Plan   (1971-1975) - M$ 48 million
- Third Malaysia Plan   (1976-1980) - M$140 million

---

[69] *Triennial Report Drainage and Irrigation Department 1961-63*, p. 115.

The area of agricultural land which benefitted from the Drainage improvement programmes amounted to approximately 404,700 ha by the end of 1980 which included major projects such as the:

(a) Western Johor Agricultural Development Project
(b) North West Selangor Integrated Development Project
(c) Trans-Perak Drainage Project
(d) Bagan Datuh Drainage Scheme

The major crops planted in the drainage schemes consist of coconut, rubber, oil palm, pineapples and other cash crops. It is further estimated that about 100,000 small-holders have benefitted from the drainage programmes. The income increase of these farmers varies from scheme to scheme, depending on the type of crops and size of farm holdings. In the case of the Western Johor Integrated Development Project, it is estimated that the annual income of the tree crops small-holders would increase from $2100 to $3700.

Operation and maintenance of completed drainage schemes are met from State funds. Drainage rates averaging from $5/- per acre ($12.36 per hectare) are collected in the gazetted drainage areas to help to offset the cost of operation and maintenance. The collected rates, in most cases are insufficient to meet the actual cost of operation and maintenance and hence to maintain a high level of service in these areas.

Problems relating to Irrigation and Drainage

Although the initial development of irrigation and drainage facilities saw a rapid increase of yields, the present yields are only moderate and quite below what the crops are capable of producing. In the case of padi this may be attributed to the lack of adequate facilities for proper water distribution and control at the farm level. Water problems mainly concern the uneven water distribution and the long time lag between the start of water delivery and the time irrigation water reaches the furthest fields. Water-logging due to inadequate drainage is also a problem which arises from the lack of adequate drainage facilities. In some cases the drainage facilities provided for evacuation of surface runoff lead to over-drainage during the drier periods. This is particularly damaging in peat and acid sulphate soils where the dehydration of peat soils and oxidation of acid sulphate would render these soils unsuitable for agricultural purposes.

Irrigation, Drainage and Rice Cultivation in West Malaysia

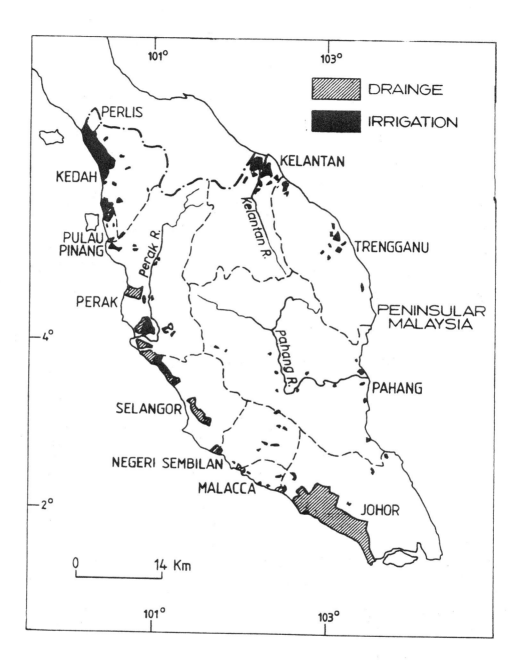

Map 5: Malaya: Major Drainage and Irrigation Areas (up to 1981)

Table 3: Areas provided with Irrigation and Agricultural Drainage Facilities, 1932-1981
(hectares)

State	Irrigation						Agricultural Drainage		
	wet season			dry season					
	1932	1957	1981	1932	1957	1981	1932	1957	1981
PERLIS	-	20,623	4,804	-	-	7,598	-	-	-
KEDAH	-	64,047	14,093	-	-	-	-	-	81
PULAU PINANG (PENANG)	-	13,561	16,755	-	2,829	15,746	-	12,175	20,877
PERAK	22,794	39,124	52,106	-	-	51,779	17,306	22,817	89,103
SELANGOR	7,848	21,300	19,264	-	-	19,214	40,027	73,616	120,777
NEGERI SEMBILAN	-	5,103	11,989	-	-	7,552	-	2,440	4,188
MELAKA	-	7,958	8,668	-	-	2,699	-	-	10,286
JOHOR	-	5,364	3,774	-	-	3,670	-	30,890	197,623
PAHANG	-	11,762	17,503	-	-	2,281	-	-	1,222
TRENGGANU	-	7,507	16,274	-	-	10,291	-	-	-
KELANTAN	-	7,185	9,806	-	-	451	-	-	8,103
MUDA	-	-	95,859	-	-	88,628	-	-	-
KADA	-	-	31,833	-	-	25,439	-	-	-
SABAH	-	-	21,590	-	-	10,643	-	-	4,775
SARAWAK	-	-	7,738	-	-	2,375	-	-	18,479
TOTAL	30,642	203,534	332,056	-	2,829	248,372	57,333	141,938	475,514

Field Water Management

The bulk of the work in irrigation development has so far been devoted to the provision of major engineering works such as storage reservoirs, pumping installations and conveyance systems while little emphasis has been given to field water management in the irrigation areas.

<u>Water rates</u>

Generally, two levels of water demand occur in padi cultivation. The first is the requirements for presaturating of the soil to enable puddling and this requires about 300 mm of water. Subsequently, water is required for replenishing losses due to evaporation, plant transpiration and percolation and this amounts to about 700 mm during the growth period. Thus, a total of about 1,000 mm are required in the field for cultivating one crop of padi. Often a substantial proportion of this water is supplied by rainfall especially during the wetter part of the year when the main crop is grown. The irrigation efficiencies for padi areas are generally low and have been estimated to be about 40 per cent. No measurements have been made so far but emphasis is currently being directed to this aspect of management.

The Irrigation Areas Ordinance was established in 1953 and this provides for the imposition of water rates in lands within gazetted irrigation areas; the rates varying with the various classes of land. The rates are collected by each of the State Governments and there is no uniformity in the amount charged by the different State Governments. Similarly, the Drainage Works Ordinance established in 1954 provides for the imposition of drainage rates and the manner in which the rate is to be determined, collected and disposed of. The rates again vary from State to State and sometimes also from area to area within a State.

Irrigation and Rice Cultivation - Continuity and Change?

Rice farming continues to play an increasingly important role in the Malaysian economy. It receives sustained attention and assistance by the government. The state has undertaken sole responsibility for investment in infrastructural construction, relying upon capital investment to achieve it. Government policy continues to "rely on technological and economic means directed at the further raising of productivity".[70]

---

[70] P.P. Courtenay, *Nearly at the Crossroads? A review of some issues raised in recent work on the Malaysian padi sector* (Canberra:

The rationale underlying the state's interest in rice production involves strategic, economic and political considerations which are not dissimilar to those that shaped colonial policy. Strategically, the government considers it vital to improve the country's self-sufficiency in rice so as to avoid serious shortages in time of war or other emergencies. Economic-political considerations arise from the fact that a substantial proportion of the population is involved in padi production and about 95 per cent of those are Malays, whose support the ruling party claims.

Consequently, since 1945 the area under padi cultivation has almost doubled while production has increased by about 60 per cent. Furthermore, in the 1920s the country imported almost three-quarters of its rice needs and rice imports have now been reduced to around 15 per cent. What are the implications of this policy for the Malays?

Studies show that the government has effectively forced rice monoculture on farmers in certain areas, notably in the Kemubu and Muda regions. Here new patterns of land use and the more intensive labour requirements of double-cropping have effectively precluded economic diversification in these regions.[71] Additionally, although mechanisation is advanced and the labour force fully commercialised in the Muda region, it has been shown that few opportunities exist for improving farmers' incomes. The intensive economic specialisation of employment and the technical demands imposed by the irrigation scheme allow no possibility of switching from rice to high-value crops.[72] This has led to a rural exodus and the concentration of land-holding accompanied by increasing mechanisation of rice. Clive Bell argues that by forcing farmers into rice monoculture so as to maximise national production, the government has increased their economic vulnerability and reduced their real incomes, in effect, sacrificing regional development to national objectives.

A.N.U. Development Studies Centre, 1983), Occasional Paper No. 33, p. 29.

[71] See for example, Mokhtar Tamin and Nik Hashim, "Kelantan, West Malaysia" in *Changes in Rice Farming in Selected Areas of Asia* (Los Banos: I.R.R.I., 1975). See also Ishak Shari and J.K. Sundaram, "Malaysia's Green Revolution in Rice Farming".

[72] See for example, Clive Bell *et al.*, *Project Evaluation in Regional Perspective*.

SELECT BIBLIOGRAPHY

Official Publications

Afifuddin Haji Omar, 1971, *Some organizational aspects of agriculture and non-agricultural growth linkages in the development of the Muda Region*, Alor Setar: MADA.

Afifuddin Haji Omar, 1977, *Irrigation structures and local peasant organisation*, Alor Setar: MADA.

Birch, Sir Ernest Woodford, 1898, *Memorandum on the subject of Irrigation*, Kuala Lumpur: Government Printer.

Ho, Nai-Kin, 1977, *Extension strategies in agricultural development - MADA's approach*, Alor Setar: MADA.

International Seminar on Irrigation Policy and Management in Southeast Asia, 1978, *Irrigation policy and management in Southeast Asia, proceedings*, Los Baños: International Rice Research Institute.

Jegatheesan, S., 1976, *Land tenure in the Muda irrigation scheme, some implications for productivity, income distribution and reform policy*, Alor Setar: MADA.

Jegatheesan, S., 1977, *The Green Revolution and the Muda Irrigation Scheme: an analysis of its impact on the size, structure and distribution of rice farmer incomes*, Alor Setar: MADA.

Mager, F.W., 1932, "A Report on Krian Irrigation Scheme", in: *Collected reports on Krian 1892-1930*, Taiping: Government of Perak.

Malaysia, 1982, *Drainage and Irrigation Department 50 Years 1932-1982*, Kuala Lumpur: Ministry of Agriculture Malaysia.

Malaysia, Bahagian Parit dan Taliayer, 1973, *Manual*, Kuala Lumpur: Government Printer.

Malaysia, Jabatan Penerangan, 1971, *Rancangan Pengayeran Kemubu*, Kuala Lumpur: Government Printer.

Malay States, Federated. Drainage and Irrigation Department, 1938-40, *Report on the Drainage and Irrigation Department of the Malay States and the Straits Settlements*, Kuala Lumpur: Government Printer (Cont. by Malaya, Malaysia).

Malay States Federated. Public Works Department, *Annual Reports 1931-1940*, Kuala Lumpur: Government Printer, (Cont. by Malaya, Malaysia).

Malay States Federated. Rice Cultivation Committee, 1931, *Report*, Kuala Lumpur: Government Printer.

Muda Agricultural Development Authority, 1972, *Muda irrigation scheme: an exercise in integrated agricultural development*, Alor Setar: MADA.

Muda Agricultural Development Authority, 1972, *Proposal; organization, structure and management policy*, Alor Setar: MADA publication no. 3.

Muda Agricultural Development Authority, 1977, *Feasibility report on tertiary irrigation facilities for intensive agricultural development in the Muda Irrigation Scheme Malaysia*, Pt 2, Vol. 1, Alor Setar: MADA.

Pinkerton, W., 1934, *Report on Investigations into the possibilities of irrigation in Kelantan*, Kuala Lumpur: FMS DID Publication no. 18, Government Printer.

Tempany, Sri Harold Augustin, 1932, *Krian Irrigation Scheme*, Taiping: Government Printer.

Books and Articles

Agarwal, M.C., 1964, "An account of the Tanjong Karang Project", *Malayan Economic Review*, Vol. 9, No. 2, pp. 64-74.

Ashton, A.V., 1940, "A Review of the Sungei Manik Padi Irrigation Scheme", *MAJ*, Vol. 28, pp. 322-9.

Bell, C. et al., 1982, *Project Evaluation in Regional Perspective: A Study of an Irrigation Project in Northwest Malaysia*, Baltimore: Johns Hopkins University Press, for the World Bank.

Bush, B.O., 1933, "Irrigation dams for small rivers", *MAJ*, Vol. 22, pp. 658-63.

Cheng Siok Hwa, 1969, "The Rice Industry of Malay: An Historical Survey", *JMBRAS*, Vol. XLII, No. 2, pp. 130-44.

Courtenay, P.P., 1983, *Nearly at the Crossroads: A review of some issues raised in recent work on the Malaysian padi sector*, Canberra: A.N.U. Development Studies Centre, Occasional Paper No. 33.

De Moubray, G.A. De C., 1936, "Sungei Manik Irrigation Scheme", *MAJ*, Vol. 24, pp.160-166.

Ding Eing Tan Soon Hai, 1963, *The Rice Industry in Malaya, 1920-1940*, Singapore: Malaya Publishing House.

Ferguson, D.S., 1954, "The Sungei Manik Irrigation Scheme", *MAJ*, Vol. 2, pp. 9-16.

Finch, F.G., 1933, "Irrigation and Drainage of padi areas", *MAJ*, Vol. 21, pp.649-57.

Fletcher, J., 1989, "Rice and Padi Marketing Management in West Malaysia, 1957-1986", *Journal of Developing Areas*, Vol. 23, No. 3, pp. 363-84.

Gittinger, J.P., 1973, "Kemubu irrigation project of Malaysia", in J.P. Gittinger, ed., *Agricultural Projects: case studies and work exercises*, Washington DC: World Bank Publication, pp. 11-51.

Goldman, R.H., 1975, "Staple Food Self-Sufficiency and the Distributive Impact of Malaysian Rice Policy", *Food Research Institute Studies*, Vol. 14, No. 3, pp.251-93.

Grist, Donald Honey, 1965, *Rice*, London: Longmans, Green & Co. (4th edition).

Hill, R.D., 1970, "Peasant rice cultivation systems with some Malaysian examples", *Geographica Polonoca*, Vol. 19, pp. 91-98.

Hill, R.D., 1977, *Rice in Malaya. A Study in Historical Geography*, Kuala Lumpur: Oxford University Press.

Ishak Shari and J.K. Sundaram, 1982, "Malaysia's Green Revolution in Rice Farming: Capital Accumulation and Technological Change in a Peasant Society", in: G.B. Hamsworth, ed., *Village-level Modernization in Southeast Asia: The Political Economy of Rice and Water*, Vancouver: University of British Columbia Press, pp. 225-54.

Kaur, Amarjit, 1980, "The Malay Peninsula in the Nineteenth Century: An Economic Survey", *Sarjana*, Vol. 4, June, pp. 69-86.

McNee, P., 1951, "Malaya: Irrigation and water supply: The Tanjong Karang Irrigation Scheme", *Corona*, Vol. 3, pp. 254-56.

Miller, J.L., 1937, "Administration of the Sungei Manik padi irrigation scheme", *MAJ*, Vol. 25, pp. 370-75.

Ouchi, T. et al., 1977, *Farmers and Villages in West Malaysia*, Tokyo: University of Tokyo Press.

Robinson, A.G., 1936, "Irrigation of riverine areas. Bota-Lambor Kanan pumping scheme, Perak river", *MAJ*, Vol. 24, pp. 524-28. [Also in *Malayan Review*, Vol. 87, 1936, pp. 65-76.]

Rowland, V.R., 1964, "Kinta River deviation", *Proceedings Institute Civil Engineers*, Vol. 27, pp. 569-587.

Rutherford, J., 1966, "Double cropping of wet padi in Penang, Malaya", *Geographical Review*, Vol. 56, pp.239-55.

Short, David Eric and James C. Jackson, 1971, "The origins of an irrigation policy in Malaya: a review of development prior to the establishment of the Drainage and Irrigation Department", *JMBRAS*, Vol. 44, Pt. 1, pp.78-103.

Short, D.E., 1980, "Irrigation Development in Malaya in the late Colonial Period 1931-57", *Malaysian Journal of Tropical Geography*, Vol. 1, No. 1, pp. 34-42.

Wilson, J.N., 1957, "The Sri Menanti, Senggarang and Muar drainage schemes, Johore", *MAJ*, Vol. 40, pp. 241-52.

Wilson, J.N., 1960, "Drainage and irrigation problems in Malaya", *Corona*, Vol. 12, pp. 302-4.

Zaharah Haji Mahmud, 1969, "The Pioneering of wet rice growing traditions in West Malaysia - a restudy with special reference to the State of Kedah", *Geographica*, Vol. 5, pp. 1-7.

Unpublished Theses, Papers

Akers, R.L. and A.I.G.S. Robertson, 1956, "Irrigation and drainage problems in Malaya", Sessional paper, Colonial Conference Institute of Civil Engineers, London.

Cheong, C.L., 1973, "Irrigation development and water management in peninsular Malaysia", Paper presented at the National Seminar on Water Management at the Farm Level, Alor Setar.

Doering, Otto Charles III, 1973, "Malaysian Rice Policy and the Muda River Irrigation Project", (unpublished Ph.D. thesis, Cornell University).

Jabatan Parit dan Tali Air, 1980, "Drainage and Irrigation Policy", Kuala Lumpur, Jabatan Parit dan Tali Air, typescript.

Short, D.E., 1971, "Some Aspects of the role of irrigation in rural development in Malaya", (unpublished Ph.D. thesis, University of Hull).

Van de Goor, 1963, "The Consumptive use of Water, the Possibilities of Double Cropping and the Irrigation Requirements for Rice in Malaya", Rome, FAO.

GLOSSARY

abdi	slave
adat	custom, tradition
air	water
ampang	weir
baru	new
bendang	wet padi field (wet rice cultivation)
bukit	hill
bumiputera	literally 'son of the soil'; indigenous
coyan (Koyan)	measure of weight or capacity equivalent to 800 *gantang* or 40 *pikul* (1 *pikul* = 100 *kati* = 133 1/3 lbs = 61.7613 kg)
damar	resin
dusun	orchard, fruit grove
gambier	(*Uncaria gambir*) ingredient in betel-quid used in tanning leather
gantang	a volumetric measure for padi, approximately equivalent to 5 1/3 lbs (1 lb = 0.453592 kg)
gelam	type of tree (*Melaleuca leucadendron*)
gotong-royong	working together; co-operation in a village
gunong (ganung)	mountain
hamba	slave; bondsman - *hamba berhutang* - a debt bondsman
hikayat	tale, chronicle
huma	dry padi cultivation
hutan	forest, jungle
kampung	group, especially of houses; hamlet or village
kangchu	district headman / 'lord of the river' applied to Chinese pepper concessionaires in Johor
kati	measure of weight equivalent to 1 1/3 lb or 0.605 kg

kerabat am/	more distant/
kerabat diraja	more close relatives of ruler
kerah	corvée, i.e. compulsory service by customary obligation
kincir	water wheel for irrigating padi fields
ladang	forest clearing for growing dry padi or other crops/dry padi cultivation
lampan	wooden tray used when washing for ore (tin, etc.)
mas	unit of currency; coin worth about 50 cents
mukim	parish, area served by a mosque (later) sub-district
negeri	state (later) country
orang asli	literally original people; the aborigines
orang berhutang	debt bondsman
orlong (or *relong*)	measure of length or of area - 240 feet or 73 metres approximately in length or 1 1/3 acres or 0.6 hectare approximately in area
padi	rice plant; unhusked rice grain as harvested
padi bukit	dry padi grown on a *ladang* (clearing)
padi paya	swamp padi - wet padi
parit	channel; irrigation canal
paya	swamp (for wet padi cultivation)
penghulu	headman of a village or (later) of an administrative sub-district - *mukim*
pikul	measure of weight; 100 kati; 133 1/3 lbs or 60.5 kg
pribumi	indigenous people
rakyat (or *raiat*, *ryot*)	subject of a ruler; peasant
relong (or *orlong*)	measure of length or of area - 240 feet or 73 metres approximately in length or 1 1/3 acres or 0.6 hectare approximately in area
sawah	wet padi fields (see *bendang* above)

serah	to hand over, deliver
sungai (sungei)	river, stream
sungai korok	man-made channel for irrigation, drainage or as a canal for movement of boats
taliair	irrigation channel to carry water from a pool or stream to the field
tongkang	a type of native craft
tugalan (padi tugalan)	padi that is planted by using a dibble
usaha	effort, diligence
waris	inheritor, heir to

APPENDIX 1

MAJOR IRRIGATION AND DRAINAGE PROJECTS IN WEST MALAYSIA

1. Muda Irrigation Project

The Muda Irrigation Project was implemented in 1966 at a cost of $228 million. The Project comprises two dams (the Muda and the Pedu) connected by an 8 km tunnel, a conveyance system equipped with flow regulating structures and an extensive internal reticulation system. It is the largest bi-annual rice producing irrigation project in Malaysia and is located on the Kedah-Perlis Plain. The area covers 126,987 ha, 72.6 per cent (or 96,000 ha) of which are under padi, and as a result of the basic irrigation infrastructure is double-cropped except for 6,000 ha which are located on high ground or isolated by natural depressions and still depend on rainfall, and are planted in the main season only. 20,304 ha of the scheme lie in Perlis.

Water sources: There are three main sources of water for farm irrigation. The rain supplies 60 per cent of the water requirement, while the Muda and Pedu storage area, with a surface area of 2,590 ha and 6,475 ha respectively, supply 30 per cent mainly for the off-season crops. The natural canals supply the remaining 10 per cent of the water requirement.

Physical Infrastructure: The irrigation and drainage facilities in the Muda area include the main canal which is 115 km in length, secondary canals and drains (2,358 km), river and coast bund (88 km) and also farm roads which total 925 km in length. The irrigation and drainage facilities end at the level of the secondary canals and drains which are boundaries of irrigation blocks. The canals and drains are too far apart (about 1.2 km to 2 km) resulting in inefficient irrigation and drainage of the area and hence are constraints to increase of productivity and incomes. Canal densities are only 10 metres per hectare. Consequently, a long-term development plan, the Tertiary System Development Plan, has been developed for the area. It is expected to be implemented within 15 years and will involve an expenditure of M$400 million.

The Muda Irrigation System

Under the Muda Irrigation Project, a system of irrigation and drainage channels was provided in the coastal plains of Kedah and Perlis to enable two crops of padi to be grown in a year. Water for this was derived from 3 sources, that is, rainfall falling directly on the padi fields, uncontrolled

Irrigation, Drainage and Rice Cultivation in West Malaysia

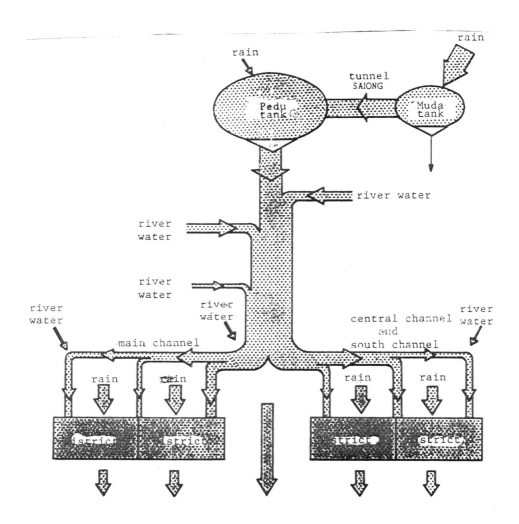

Figure 3: The Distribution of Water

flow from rivers, and irrigation water from the two reservoirs Muda and Pedu. These water sources are integrated into a single control system suitable for double-cropping of padi.

2. Kemubu Irrigation Project

The Kemubu Irrigation Project was designed to irrigate 18,350 ha of padi land for double cropping by pumping from a major river. The project was implemented in 1960 and commissioned in 1972 for the main season cropping of agricultural land. It is administered by the Kemubu Agricultural Development Authority or KADA. KADA is responsible for the maintenance of five irrigation schemes, namely, the Kemubu Irrigation Scheme (19,000 hectares), Lemal Irrigation Scheme (9,300 hectares), Pasir Mas Irrigation Scheme (2,100 hectares), Salor Irrigation Scheme (850 hectares) and the Alor Pasir Irrigation Scheme (230 hectares). The total area of land under the jurisdiction of KADA is about 60,438 hectares of which 31,480 hectares are suitable for double-cropping of padi. The Kemubu and Salor Irrigation Schemes are situated on the east bank of the Kelantan River while the Lemal, Pasir Mas and Alor Pasir Irrigation Schemes are on its west bank. The Kemubu Irrigation Scheme is the largest of the five schemes. The money spent on engineering works of this scheme amounted to $75 million, of which $30 million were loaned by the World Bank. The Lemal, Alor Pasir, Pasir Mas and Salor Irrigation Schemes were started in the mid 1960s when a total of $4.7 million was spent for infrastructure in the respective areas.

3. Besut Irrigation Project

The Besut Irrigation Project, a component of the Besut Agricultural Development Project, was implemented in 1972 at a cost of $16 million. The Project is located on the right bank of Sungai Besut in the northern part of the State of Trengganu and comprises a river barrage, conveyance and internal reticulation systems and various types of irrigation structures. The project seeks to provide irrigation facilities to 5,100 ha of padi land for double cropping. By 1974 the entire area was planted with the main season crop and 1,200 ha of the off season crop. In 1976 a sum of $10 million was allocated for the provision of additional drainage and irrigation facilities for some 1,200 ha.

4. National Small-Scale Irrigation Project

The main objective of the National Small-Scale Irrigation Project is to raise the productivity and incomes of over 60,000 padi farming families whose padi lands are located outside the large scale irrigation areas. The project covers some 195 small padi areas scattered throughout the country. The project commenced in 1977 at a cost of $244 million.

5. Tanjung Karang Irrigation Project

The Tanjung Karang Irrigation Project is a component of the Northwest Selangor Integrated Agricultural Development Project which was implemented in 1977 at a total project cost of $149 million. The project comprises the rehabilitation of major structures and canals and construction of tertiary irrigation, drainage and farm road networks. The project which extends over 20,000 ha aims to improve the productivity and income of some 17,000 farm families by increasing the average annual farm income from the present $3,500 to $4,400 within 10 years of project implementation. The project is located in a region bordering the northwest coast of the State of Selangor.

6. Krian-Sungai Manik Irrigation Project

The Krian-Sungai Manik Irrigation Project is a major component of the Krian-Sungai Manik Integrated Agricultural Development Project. The main objective of the Project is to raise the productivity and incomes of about 24,000 padi farming families in the Krian Irrigation area (236,000 ha) in the northwest corner of the State of Perak and in the Sungai Manik Irrigation area (7,000 ha) located in southern Perak. The project was implemented in 1978 and comprises the rehabilitation and improvement of the existing primary irrigation and drainage system and the construction of tertiary and quaternary irrigation and drainage channels and access roads. The project is expected to increase the present average annual farm income of $2,500 to $3,300 within a decade of project implementation.

7. Western Johor Agricultural Development Project

The Western Johor Agricultural Development Project is situated on the west coast of the State of Johor. The project covers the provision of drainage facilities, roads, embankments and agricultural services to 146,800 ha of existing and potential agricultural land. The project commenced in 1974 with a total estimated cost of $233 million. The project aimed to create 12,000 full time jobs within 10 years of implementation and increase the average annual farm income from the present $2,100 to $3,700 within this time.

8. Sabak Bernam-Kuala Selangor Drainage Project

The Sabak Bernam-Kuala Selangor Drainage Project is the drainage component of the Northwest Selangor Integrated Agricultural Development Project which was implemented in 1978 at a total cost of $149 million. The project is located in a region bordering the northwest coast of the State of Selangor and comprises the improvement of existing drainage, flood protection works, and the provision of internal drainage networks. The project aimed to benefit 20,000 smallholders' families cultivating some 43,000 ha of tree crops, and to raise the income of these families through intercropping coconut with cocoa or coffee and replanting with high yielding varieties. The average annual farm income of smallholders was projected to increase within 10 years from the present $2,200 to between $5,600 and $9,100 depending on the choice of primary crops and intercrops.